I0037225

This report contains the collective view: [...] experts and does not necessarily represent the decisions or the stated policy of the United Nations Environment Programme, the International Labour Organization, or the World Health Organization.

Environmental Health Criteria 219

FUMONISIN B_1

First draft prepared by Professor W.F.O. Marasas (Medical Research Council, Tygerberg, South Africa), Professor J.D. Miller (Carleton University, Ottawa, Canada), Dr R.T. Riley (US Department of Agriculture, Athens, USA) and Dr A. Visconti (National Research Council, Bari, Italy)

Published under the joint sponsorship of the United Nations Environment Programme, the International Labour Organization, and the World Health Organization, and produced within the framework of the Inter-Organization Programme for the Sound Management of Chemicals.

World Health Organization
Geneva, 2000

The **International Programme on Chemical Safety (IPCS)**, established in 1980, is a joint venture of the United Nations Environment Programme (UNEP), the International Labour Organisation (ILO) and the World Health Organization (WHO). The overall objectives of the IPCS are to establish the scientific basis for assessment of the risk to human health and the environment from exposure to chemicals, through international peer review processes, as a prerequisite for the promotion of chemical safety, and to provide technical assistance in strengthening national capacities for the sound management of chemicals.

The **Inter-Organization Programme for the Sound Management of Chemicals (IOMC)** was established in 1995 by UNEP, ILO, the Food and Agriculture Organization of the United Nations, WHO, the United Nations Industrial Development Organization and the Organisation for Economic Co-operation and Development (Participating Organizations), following recommendations made by the 1992 UN Conference on Environment and Development to strengthen cooperation and increase coordination in the field of chemical safety. The purpose of the IOMC is to promote coordination of the policies and activities pursued by the Participating Organizations, jointly or separately, to achieve the sound management of chemicals in relation to human health and the environment.

WHO Library Cataloguing-in-Publication Data

Fumonisin B$_1$.

(Environmental health criteria ; 219)

1.Carboxylic acids - toxicity   2.Food contamination   3.Environmental exposure
4.Risk assessment   I.Series

ISBN 92 4 157219 1   (NLM Classification: QD 341.P5)
ISSN 0250-863X

Computer typesetting by I. Xavier Lourduraj, Chennai, India

Printed in Finland
2000/13161 – Vammala – 5000

CONTENTS

ENVIRONMENTAL HEALTH CRITERIA FOR
FUMONISIN B₁

NOTE TO READERS OF THE CRITERIA MONOGRAPHS

Every effort has been made to present information in the criteria monographs as accurately as possible without unduly delaying their publication. In the interest of all users of the Environmental Health Criteria monographs, readers are requested to communicate any errors that may have occurred to the Director of the International Programme on Chemical Safety, World Health Organization, Geneva, Switzerland, in order that they may be included in corrigenda.

*   *   *

A detailed data profile and a legal file can be obtained from the International Register of Potentially Toxic Chemicals, Case postale 356, 1219 Châtelaine, Geneva, Switzerland (telephone no. + 41 22 - 9799111, fax no. + 41 22 - 7973460, E-mail irptc@unep.ch).

*   *   *

This publication was made possible by grant number 5 U01 ES02617-15 from the National Institute of Environmental Health Sciences, National Institutes of Health, USA, and by financial support from the European Commission.

Environmental Health Criteria

PREAMBLE

Objectives

In 1973 the WHO Environmental Health Criteria Programme was initiated with the following objectives:

(i)   to assess information on the relationship between exposure to environmental pollutants and human health, and to provide guidelines for setting exposure limits;

(ii)   to identify new or potential pollutants;

(iii)   to identify gaps in knowledge concerning the health effects of pollutants;

(iv)   to promote the harmonization of toxicological and epidemiological methods in order to have internationally comparable results.

The first Environmental Health Criteria (EHC) monograph, on mercury, was published in 1976 and since that time an ever-increasing number of assessments of chemicals and of physical effects have been produced. In addition, many EHC monographs have been devoted to evaluating toxicological methodology, e.g. for genetic, neurotoxic, teratogenic and nephrotoxic effects. Other publications have been concerned with epidemiological guidelines, evaluation of short-term tests for carcinogens, biomarkers, effects on the elderly and so forth.

Since its inauguration the EHC Programme has widened its scope, and the importance of environmental effects, in addition to health effects, has been increasingly emphasized in the total evaluation of chemicals.

The original impetus for the Programme came from World Health Assembly resolutions and the recommendations of the 1972 UN Conference on the Human Environment. Subsequently the work became an integral part of the International Programme on Chemical Safety (IPCS), a cooperative programme of UNEP, ILO and WHO. In this manner, with the strong support of the new partners, the importance of occupational health and environmental effects was fully recognized. The EHC monographs have become widely established, used and recognized throughout the world.

The recommendations of the 1992 UN Conference on Environment and Development and the subsequent establishment of the Intergovernmental Forum on Chemical Safety with the priorities for action in the six programme areas of Chapter 19, Agenda 21, all lend further weight to the need for EHC assessments of the risks of chemicals.

Scope

The criteria monographs are intended to provide critical reviews on the effect on human health and the environment of chemicals and of combinations of chemicals and physical and biological agents. As such, they include and review studies that are of direct relevance for the evaluation. However, they do not describe *every* study carried out. Worldwide data are used and are quoted from original studies, not from abstracts or reviews. Both published and unpublished reports are considered and it is incumbent on the authors to assess all the articles cited in the references. Preference is always given to published data. Unpublished data are used only when relevant published data are absent or when they are pivotal to the risk assessment. A detailed policy statement is available that describes the procedures used for unpublished proprietary data so that this information can be used in the evaluation without compromising its confidential nature (WHO (1990) Revised Guidelines for the Preparation of Environmental Health Criteria Monographs. PCS/90.69, Geneva, World Health Organization).

In the evaluation of human health risks, sound human data, whenever available, are preferred to animal data. Animal and *in vitro* studies provide support and are used mainly to supply evidence missing from human studies. It is mandatory that research on human subjects is conducted in full accord with ethical principles, including the provisions of the Helsinki Declaration.

The EHC monographs are intended to assist national and international authorities in making risk assessments and subsequent risk management decisions. They represent a thorough evaluation of risks and are not, in any sense, recommendations for regulation or standard setting. These latter are the exclusive purview of national and regional governments.

Content

The layout of EHC monographs for chemicals is outlined below.

* Summary – a review of the salient facts and the risk evaluation of the chemical
* Identity – physical and chemical properties, analytical methods
* Sources of exposure
* Environmental transport, distribution and transformation
* Environmental levels and human exposure
* Kinetics and metabolism in laboratory animals and humans
* Effects on laboratory mammals and *in vitro* test systems
* Effects on humans
* Effects on other organisms in the laboratory and field
* Evaluation of human health risks and effects on the environment
* Conclusions and recommendations for protection of human health and the environment
* Further research
* Previous evaluations by international bodies, e.g. IARC, JECFA, JMPR

Selection of chemicals

Since the inception of the EHC Programme, the IPCS has organized meetings of scientists to establish lists of priority chemicals for subsequent evaluation. Such meetings have been held in Ispra, Italy, 1980; Oxford, United Kingdom, 1984; Berlin, Germany, 1987; and North Carolina, USA, 1995. The selection of chemicals has been based on the following criteria: the existence of scientific evidence that the substance presents a hazard to human health and/or the environment; the possible use, persistence, accumulation or degradation of the substance shows that there may be significant human or environmental exposure; the size and nature of populations at risk (both human and other species) and risks for environment; international concern, i.e. the substance is of major interest to several countries; adequate data on the hazards are available.

If an EHC monograph is proposed for a chemical not on the priority list, the IPCS Secretariat consults with the Cooperating Organizations and all the Participating Institutions before embarking on the preparation of the monograph.

Procedures

The order of procedures that result in the publication of an EHC monograph is shown in the flow chart on p. xii. A designated staff member of IPCS, responsible for the scientific quality of the document, serves as Responsible Officer (RO). The IPCS Editor is responsible for layout and language. The first draft, prepared by consultants or, more usually, staff from an IPCS Participating Institution, is based initially on data provided from the International Register of Potentially Toxic Chemicals, and reference data bases such as Medline and Toxline.

The draft document, when received by the RO, may require an initial review by a small panel of experts to determine its scientific quality and objectivity. Once the RO finds the document acceptable as a first draft, it is distributed, in its unedited form, to well over 150 EHC contact points throughout the world who are asked to comment on its completeness and accuracy and, where necessary, provide additional material. The contact points, usually designated by governments, may be Participating Institutions, IPCS Focal Points, or individual scientists known for their particular expertise. Generally some four months are allowed before the comments are considered by the RO and author(s). A second draft incorporating comments received and approved by the Director, IPCS, is then distributed to Task Group members, who carry out the peer review, at least six weeks before their meeting.

The Task Group members serve as individual scientists, not as representatives of any organization, government or industry. Their function is to evaluate the accuracy, significance and relevance of the information in the document and to assess the health and environmental risks from exposure to the chemical. A summary and recommendations for further research and improved safety aspects are also required. The composition of the Task Group is dictated by the range of expertise required for the subject of the meeting and by the need for a balanced geographical distribution.

EHC PREPARATION FLOW CHART

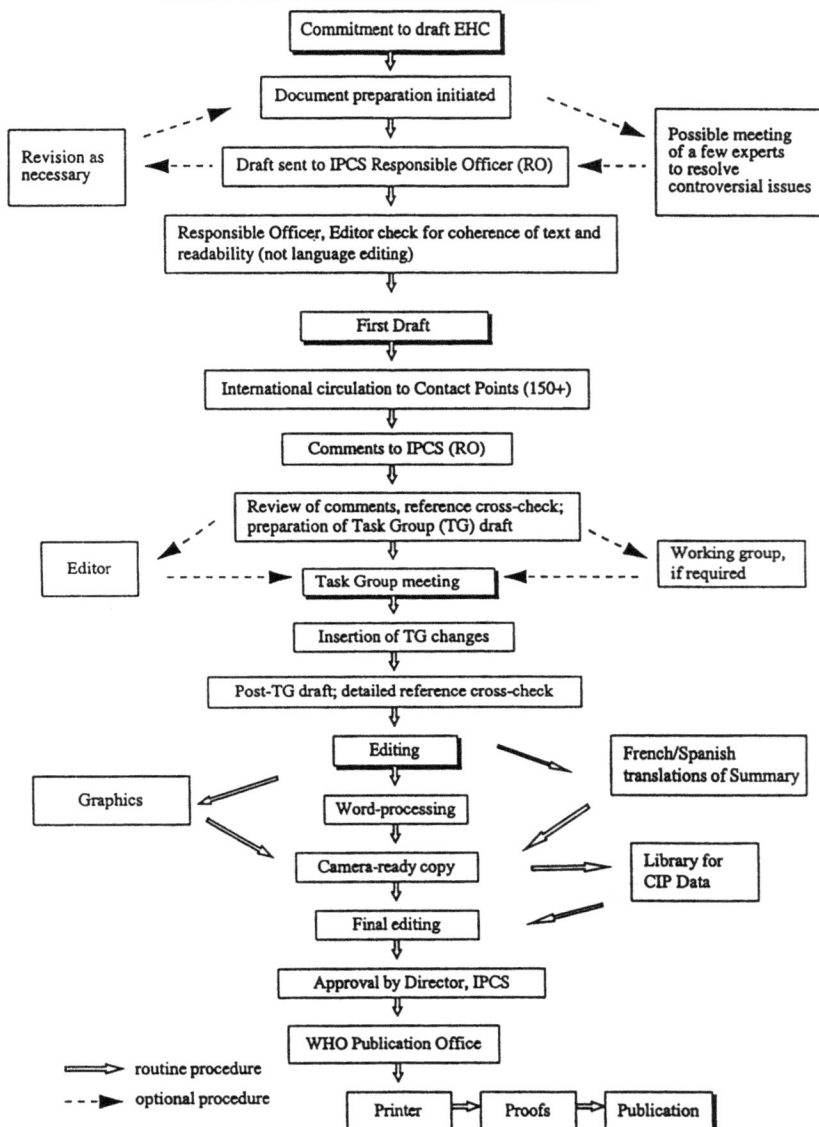

Commitment to draft EHC

⇩

Document preparation initiated

⇩

Draft sent to IPCS Responsible Officer (RO)

Revision as necessary

Possible meeting of a few experts to resolve controversial issues

Responsible Officer, Editor check for coherence of text and readability (not language editing)

⇩

First Draft

⇩

International circulation to Contact Points (150+)

⇩

Comments to IPCS (RO)

⇩

Review of comments, reference cross-check; preparation of Task Group (TG) draft

Editor

Working group, if required

Task Group meeting

⇩

Insertion of TG changes

⇩

Post-TG draft; detailed reference cross-check

Editing

French/Spanish translations of Summary

Graphics

Word-processing

Camera-ready copy

Library for CIP Data

Final editing

⇩

Approval by Director, IPCS

⇩

WHO Publication Office

⇩

⟹ routine procedure

- - -▶ optional procedure

Printer ⇨ Proofs ⇨ Publication

xii

The three cooperating organizations of the IPCS recognize the important role played by nongovernmental organizations. Representatives from relevant national and international associations may be invited to join the Task Group as observers. Although observers may provide a valuable contribution to the process, they can only speak at the invitation of the Chairperson. Observers do not participate in the final evaluation of the chemical; this is the sole responsibility of the Task Group members. When the Task Group considers it to be appropriate, it may meet *in camera*.

All individuals who as authors, consultants or advisers participate in the preparation of the EHC monograph must, in addition to serving in their personal capacity as scientists, inform the RO if at any time a conflict of interest, whether actual or potential, could be perceived in their work. They are required to sign a conflict of interest statement. Such a procedure ensures the transparency and probity of the process.

When the Task Group has completed its review and the RO is satisfied as to the scientific correctness and completeness of the document, it then goes for language editing, reference checking and preparation of camera-ready copy. After approval by the Director, IPCS, the monograph is submitted to the WHO Office of Publications for printing. At this time a copy of the final draft is sent to the Chairperson and Rapporteur of the Task Group to check for any errors.

It is accepted that the following criteria should initiate the updating of an EHC monograph: new data are available that would substantially change the evaluation; there is public concern for health or environmental effects of the agent because of greater exposure; an appreciable time period has elapsed since the last evaluation.

All Participating Institutions are informed, through the EHC progress report, of the authors and institutions proposed for the drafting of the documents. A comprehensive file of all comments received on drafts of each EHC monograph is maintained and is available on request. The Chairpersons of Task Groups are briefed before each meeting on their role and responsibility in ensuring that these rules are followed.

WHO TASK GROUP ON ENVIRONMENTAL HEALTH CRITERIA FOR FUMONISIN B₁

Members

Dr R.V. Bhat, Food and Drug Toxicology Research Centre, National Institute of Nutrition, Indian Council of Medical Research, Hyderabad, India

Dr M. Hirose, Division of Pathology, Biological Research Centre, National Institute of Health Sciences, Tokyo, Japan

Dr P.C. Howard, Division of Biochemical Toxicology, National Center for Toxicology Research, US Food and Drug Administration, Jefferson, Arkansas, USA

Dr S. Humphreys, Center for Food Safety and Applied Nutrition, US Food and Drug Administration, Washington DC, USA

Professor M. Kirsch-Volders, Laboratory for Cellular Genetics, Brussels, Belgium (*Chairman*)

Professor W.F.O. Marasas, Medical Research Council, Tygerberg, South Africa

Professor J.D. Miller, Department of Chemistry, Carleton University, Ottawa, Ontario, Canada

Dr J.H. Olsen, Institute of Cancer Epidemiology, Danish Cancer Society, Copenhagen, Denmark

Dr R. Plestina, Toxicology Unit, Institute for Medical Research and Occupational Health, Zagreb, Croatia

Dr R.T. Riley, Agricultural Research Service, US Department of Agriculture, Athens, USA

Dr A. Visconti, Institute for Toxins and Mycotoxins of Plant Parasites, National Research Council, Bari, Italy (*Vice-Chairman*)

Secretariat

Dr A. Aitio, International Programme on Chemical Safety, World Health Organization, Geneva, Switzerland (*Joint Secretary*)

Mr Y. Hayashi, International Programme on Chemical Safety, World Health Organization, Geneva, Switzerland (*Joint Secretary*)

Dr J.M. Rice, International Agency for Research on Cancer, Lyon, France

WHO TASK GROUP ON ENVIRONMENTAL HEALTH CRITERIA FOR FUMONISIN B$_1$

A WHO Task Group on Environmental Health Criteria for Fumonisin B$_1$ met at the World Health Organization, Geneva, Switzerland from 10 to 14 May 1999. Dr M. Younes, Acting Coordinator, Programme for the Promotion of Chemical Safety, opened the meeting and welcomed the participants on behalf of the IPCS and its three cooperating organizations (UNEP/ILO/WHO). The Task Group reviewed and revised the draft monograph and made an evaluation of the risks for human health and the environment from exposure to fumonisin B$_1$.

Professor W.F.O. Marasas, Professor J.D. Miller, Dr R.T. Riley and Dr A. Visconti prepared the first draft of this monograph. The second draft incorporated comments received following the circulation of the first draft to the IPCS Contact Points for Environmental Health Criteria monographs.

Dr A. Aitio, Mr Y. Hayashi and Dr P. Jenkins of the IPCS Central Unit were responsible for the overall scientific content and technical editing, respectively.

The efforts of all who helped in the preparation and finalization of the monograph are gratefully acknowledged.

* * *

Financial support for this Task Group was provided by the US Food and Drug Administration as part of its contributions to the IPCS.

ABBREVIATIONS

2-AAF	2-acetylaminofluorene
AAL-toxin	*Alternaria alternata lycopersici* toxin
AMP	adenosine monophosphate
AP	aminopentol
CV	coefficient of variation
CZE	capillary zone electrophoresis
DEN	diethylnitrosamine
DNA	deoxyribonucleic acid
EDL	effective dose level
EGF	epidermal growth factor
ELEM	equine leukoencephalomalacia
ELISA	enzyme-linked immunosorbent assay
FA, FAK	fumonisin A, fumonisin AK
FB	fumonisin B
FC	fumonisin C
FP	fumonisin P
GC	gas chromatography
GGT	gamma-glutamyltranspeptidase
HPLC	high-performance liquid chromatography
IC_{50}	median inhibitory concentration
LC_{50}	median lethal concentration
LPS	lipopolysaccharide
MAPK	mitogen-activated protein kinase
MME	monomethyl ester
MS	mass spectrometry
NADH	reduced nicotinamide adenine dinucleotide
NADPH	reduced nicotinamide adenine dinucleotide phosphate
NCTR	National Center for Toxicological Research (USA)
NMBA	*N*-methylbenzylnitrosamine
NOEL	no-observed-effect level

NTD	neural tube defect
NTP	National Toxicology Program (USA)
OPA	*o*-phthaldialdehyde
PDI	probable daily intake
PFC	plaque-forming cell
PGST	placental glutathione *S*-transferase
PIM	pulmonary intravascular/interstitial macrophage
PKC	protein kinase C
PPE	porcine pulmonary oedema
PUFA	polyunsaturated fatty acid
Sa/So	sphinganine/sphingosine
TCA	tricarbalyllic acid moiety
TLC	thin-layer chromatography
TNF-α	tumour necrosis factor-α

INTRODUCTION

In this document, the fungus previously referred to as *Fusarium moniliforme* Sheldon, is referred to as *Fusarium verticillioides* (Sacc.) Nirenberg in accordance with a decision taken at the 8th International *Fusarium* Workshop held at CABI BioScience, Egham, United Kingdom, 17-20 August 1998.

This monograph focuses on fumonisin B_1, the most abundant naturally occurring fumonisin. Some information is also given on fumonisins B_2 and B_3, which frequently occur with FB_1, both in culture material and in naturally contaminated samples.

1. SUMMARY, EVALUATION AND RECOMMENDATIONS

1.1 Summary

1.1.1 Identity, physical and chemical properties, and analytical methods

Fumonisin B_1 (FB_1) has the empirical formula $C_{34}H_{59}NO_{15}$ and is the diester of propane-1,2,3-tricarboxylic acid and 2-amino-12,16-dimethyl-3,5,10,14,15-pentahydroxyeicosane (relative molecular mass: 721). It is the most prevalent of fumonisins, a family of toxins with at least 15 identified members. The pure substance is a white hygroscopic powder, which is soluble in water, acetonitrile-water or methanol, is stable in acetonitrile-water (1:1), is unstable in methanol, and is stable at food processing temperature and to light.

Several analytical methods have been reported, including thin-layer chromatography (TLC) and liquid chromatographic (LC), mass spectroscopic (MS), post-hydrolysis gas chromatographic and immunochemical methods, although the majority of studies have been performed using LC analysis of a fluorescent derivative.

1.1.2 Sources of human exposure

FB_1 is produced by several *Fusarium* species, mainly by *Fusarium verticillioides* (Sacc.) Nirenberg (= *Fusarium moniliforme* Sheldon), which is one of the most common fungi associated with maize worldwide. Significant accumulation of FB_1 in maize occurs when weather conditions favour Fusarium kernel rot.

1.1.3 Environmental transport, distribution and transformation

There is evidence that fumonisins can be metabolized by some soil microorganisms. However, little is known about the environmental fate of fumonisins after they are either excreted or processed.

1.1.4 Environmental levels and human exposure

FB_1 has been detected in maize and maize-based products worldwide at mg/kg levels, sometimes in combination with other mycotoxins. Concentrations at mg/kg levels have also been reported in food for human consumption. Dry milling of maize results in the distribution of fumonisin into the bran, germ and flour. In experimental wet milling, fumonisin was detected in steep water, gluten, fibre and germ, but not in the starch. FB_1 is stable in maize and polenta, whereas it is hydrolysed in nixtamalized maize-based foods, i.e. foods processed with hot alkali solutions.

FB_1 is not present in milk, meat or eggs from animals fed grain containing FB_1 at levels that would not affect the health of the animals. Human exposure estimates for the USA, Canada, Switzerland, the Netherlands and the Transkei (South Africa) ranged from 0.017 to 440 µg/kg body weight per day. No data on occupational inhalation exposure are available.

1.1.5 Kinetics and metabolism in animals

There have been no reports on the kinetics or metabolism of FB_1 in humans. In experimental animals it is poorly absorbed when dosed orally, is rapidly eliminated from circulation and is recovered unmetabolized in faeces. Biliary excretion is important, and small amounts are excreted in urine. It can be degraded to partially hydrolysed FB_1 in the gut of non-human primates and some ruminants. A small amount is retained in the liver and kidney.

1.1.6 Effects on animals and in vitro test systems

FB_1 is hepatotoxic in all animal species tested including mice, rats, equids, rabbits, pigs and non-human primates. With the exception of Syrian hamsters, embryotoxicity or teratogenicity is only observed concurrent with or subsequent to maternal toxicity. Fumonisins are nephrotoxic in pigs, rats, sheep, mice and rabbits. In rats and rabbits, renal toxicity occurs at lower doses than hepatotoxicity. Fumonisins are known to be the cause of equine leukoencephalomalacia and porcine pulmonary oedema syndrome, both associated with the consumption of maize-based feeds. Limited information on immunological properties of FB_1 is available. It was hepatocarcinogenic to male rats in one strain and nephrocarcinogenic in another strain at the same dose levels (50 mg/kg diet), and was hepatocarcinogenic at 50 mg/kg diet in female mice. There appears to be a correlation between organ toxicity and cancer development. FB_1 was the first specific inhibitor of *de novo* sphingolipid metabolism to be discovered and is currently widely used to study the role of sphingolipids in cellular regulation. FB_1 inhibits cell growth and causes accumulation of free sphingoid bases and alteration of lipid metabolism in animals, plants and some yeasts. It did not induce gene mutations in bacteria or unscheduled DNA synthesis in primary rat hepatocytes, but induced a dose-dependent increase in chromosomal aberrations at low concentration levels in one study on primary rat hepatocytes.

1.1.7 Effects on humans

There are no confirmed records of acute fumonisin toxicity in humans. Available correlation studies from the Transkei, South Africa, suggest a link between dietary fumonisin exposure and oesophageal cancer. This was observed where relatively high fumonisin exposure has been demonstrated and where environmental conditions promote fumonisin accumulation in maize, which is the staple diet. Correlation studies are also available from China. However, no clear picture on the relationship between either fumonisin or *F. verticillioides* contamination and oesophageal cancer emerged. Owing to the absence of fumonisin exposure data, no conclusion can be drawn from a case control study of males in Italy showing an association between maize intake and upper gastrointestinal tract cancer among subjects with high alcohol consumption.

There are no validated biomarkers for human exposure to FB_1.

1.1.8 Effects on other organisms in the laboratory

FB_1 inhibits cell growth and causes accumulation of free sphingoid bases and alteration of lipid metabolism in *Saccharomyces cerevisiae*.

FB_1 is phytotoxic, damages cell membranes and reduces chlorophyll synthesis. It also disrupts the biosynthesis of sphingolipids in plants and may play a role in the pathogenicity of maize by fumonisin-producing *Fusarium* species.

3

1.2 Evaluation of human health risks

1.2.1 Exposure

Human exposure as demonstrated by the occurrence of FB_1 in maize intended for human consumption is common worldwide. There are considerable differences in the extent of human exposure between different maize-growing regions. This is most evident when comparing fully developed and developing countries. For example, although FB_1 can occur in maize products in the USA, Canada and western Europe, human consumption of those products is modest. In parts of Africa, South-Central America and Asia, some populations consume a high percentage of their calories as maize meal where FB_1 contamination may be high (see Appendix 2). Maize contaminated naturally by FB_1 can be simultaneously contaminated with other *F. verticillioides* or *F. proliferatum* toxins or with other agriculturally important toxins including deoxynivalenol, zearalenone, aflatoxin and ochratoxin.

FB_1 is stable to food processing methods used in North America and western Europe. Treating maize with base and/or water washing effectively lowers the FB_1 concentrations. However, its hepatotoxicity and/or nephrotoxicity in experimental animals are still evident. Little is known about how food processing techniques used in the developing world affect FB_1 in maize products.

1.2.2 Hazard identification

The causal role of FB_1 exposure in the disease equine leukoencephalomalacia has been established. Large-scale outbreaks of this fatal disease occurred in the USA during the 19th century and as recently as 1989-1990. The causal role of FB_1 exposure in the fatal disease porcine pulmonary oedema has been established. As observed in pregnant females, low exposures to FB_1 are fatal to rabbits. Exposure has been demonstrated to result in renal toxicity and causes hepatotoxicity in all animal species studied, including non-human primates. FB_1 exposure causes hypercholesterolaemia in several animal species, including non-human primates. There is good evidence for altered lipid metabolism in the animal diseases associated with FB_1 exposure. Disruption of sphingolipid metabolism is evident either before or concurrent with *in vitro* and *in vivo* toxicity. The use of fumonisins as tools to study the function of sphingolipids has revealed that sphingolipids are required for cell growth and affect signalling molecules in several pathways, leading to apoptotic and necrotic cell death, cellular differentiation and altered immune responses. Altered lipid metabolism and changes in the activity and/or expression of key enzymes responsible for normal cell cycle progress appear to be common factors following exposure to FB_1. FB_1 is not a developmental toxin to rat, mouse or rabbit. It induces fetotoxicity in Syrian hamster at high doses without maternal toxicity.

The carcinogenicity of FB_1 in rodents varies between species, strains and sex. The only study with $B6C3F_1$ mice indicated that FB_1 was hepatocarcinogenic to females at 50 mg/kg in the diet. Primary hepatocellular carcinomas and cholangial carcinomas were induced in male BD IX rats fed diets at 50 mg FB_1/kg for up to 26 months. Renal tubule adenomas and carcinomas were detected in male F344/N Nctr rats fed 50 mg FB_1/kg. There appears to be a correlation between organ toxicity and cancer development.

A limited number of genotoxicity studies are available. FB_1 was not mutagenic in bacterial assays. In *in vitro* mammalian cells, unscheduled DNA synthesis was not detected but FB_1 caused chromosomal breaks in rat hepatocytes in one study. Other studies have shown that FB_1 causes increased lipid peroxidation *in vivo* and *in vitro*. It is possible that chromosome-breaking effects and lipid peroxidation are causally related.

FB_1 levels above 100 mg/kg, which have been reported in maize consumed by humans in Africa and China, would probably cause leukoencephalomalacia, pulmonary oedema syndrome or cancer if fed to horses, pigs and rats or mice, respectively. Despite these cases of very high human exposure, there are no confirmed records of acute fumonisin toxicity in humans. Available correlation studies from the Transkei, South Africa, suggest a link between dietary fumonisin exposure and oesophageal cancer. Elevated rates of oesophageal cancer have been observed where relatively high fumonisin exposure has been demonstrated and where environmental conditions promote the accumulation of fumonisin in maize, which is the staple diet.

One case-control study in males from Italy found an association between maize intake and cancers of the upper digestive tract,

5

including oesophageal cancer, among subjects with high alcohol consumption. There were no data on fumonisin exposure.

1.2.3 Dose-response assessment

The lowest dose of FB_1 that induced hepatocarcinomas in experimental animals was 50 mg/kg diet in male BD IX rats and female $B6C3F_1$/Nctr mice; no cancer induction was observed at 25 or 15 mg/kg diet, respectively. In each case, indications of hepatotoxicity or lipid alterations were noted at the same or lower doses in studies with these same rat and mouse strains. The lowest dose of FB_1 that induced renal carcinomas in the male F344/N Nctr rats was 50 mg/kg diet; no cancer induction was observed at 15 mg/kg diet. Renal tubular apoptosis and cell proliferation, as well as tissue and urinary sphingolipid changes, occurred at lower doses than those required for the induction of cancer in these studies.

No data are available to assess quantitatively the relationship between exposure to FB_1 and possible effects in humans.

1.2.4 Risk characterization

FB_1 is carcinogenic in mice and rats and induces fatal diseases in pigs and horses at levels of exposure that humans encounter. The Task Group was not in a position to perform a quantitative estimation of the human health risks, but considered that such an estimation is urgently needed.

1.3 Recommendations for protection of human health

a) Limits for human dietary exposure should be established. Special consideration should be given to populations consuming a high percentage of their calories as maize meal.

b) Measures should be taken to limit fumonisin exposure and maize contamination by:

- planting alternative crops in areas where maize is not well adapted;
- developing maize resistant to Fusarium kernel rot;
- practising better crop management;
- segregating mouldy kernels.

c) Early awareness of potential food contamination should be increased by improving communication between veterinarians and public health officials on outbreaks of mycotoxicoses in domestic animals.

d) A robust, low-cost and simple screening method for the detection of fumonisin contamination in maize should be developed.

2. IDENTITY, PHYSICAL AND CHEMICAL PROPERTIES, AND ANALYTICAL METHODS

2.1 Identity

Common name: Fumonisin B_1 (FB_1)

Chemical formula: $C_{34}H_{59}NO_{15}$

Chemical structure:

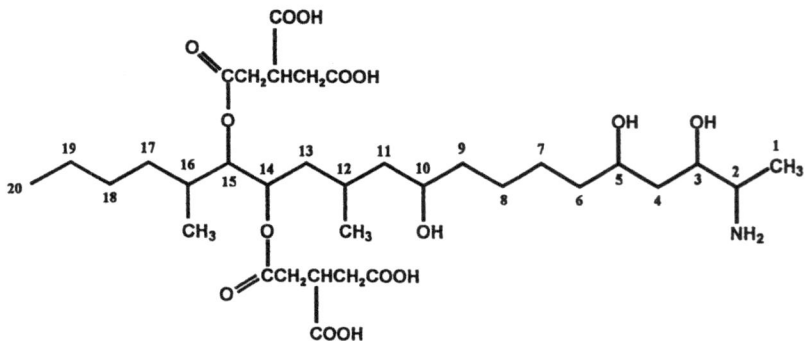

Relative molecular mass: 721

CAS Name: 1,2,3-Propanetricarboxylic acid, 1,1′-[1-(12-amino-4,9,11-trihydroxy-2-methyl-tridecyl)-2-(1-methylpentyl)-1,2-ethane-diyl] ester

IUPAC name: None

CAS registry number: 116355-83-0

RTECS No.:   TZ 8350000

8

Synonym: Macrofusine

At least 15 different fumonisins have so far been reported and other minor metabolites have been identified, although most of them have not been shown to occur naturally. They have been grouped into four main categories (Plattner, 1995; Abbas & Shier, 1997; Musser & Plattner, 1997): FA_1, FA_2, FA_3, FAK_1; FB_1, FB_2, FB_3, FB_4; FC_1, FC_2, FC_3, FC_4; FP_1, FP_2 and FP_3. FB_2, FB_3 and FB_4 differ from FB_1 in that they lack hydroxyl groups present in FB_1; FA_1, FA_2 and FA_3 are like FB_1, FB_2 and FB_3, but are N-acetylated; FAK_1 is like FA_1 but is 15-keto functionalized; FCs are like FBs but lack the methyl group adjacent to the amino group; FPs have a 3-hydroxypyridium group instead of the amine group in the FBs. This monograph will focus mainly on FB_1, the most abundant of the naturally occurring fumonisins.

2.2 Physical and chemical properties of the pure substance

Physical state: White hygroscopic powder

Melting point: Not known (has not been crystallized)

Optical rotation: Not known

Spectroscopy: Mass spectral and nuclear magnetic resonance data are given in Bezuidenhout et al. (1988), Laurent et al. (1989a) and Savard & Blackwell (1994)

Solubility: Soluble in water to at least to 20 mg/ml (US NTP, 1999); soluble in methanol, acetonitrile-water.

n-Octanol/water par-   1.84 (Norred et al., 1997)
tition coefficient (log P):

Stability: Stable in acetonitrile-water (1:1) for up to 6 months at 25 °C; unstable in methanol (25% or 35% concentration decrease after

3 or 6 weeks at 25 °C, respectively), giving
rise to monomethyl or dimethyl esters
(Gelderblom et al., 1992a; Visconti et al.,
1994); stable in methanol up to 6 weeks at
−18 °C (Visconti et al., 1994); stable at
78 °C for 16 h in buffer solutions at pH
between 3.5 and 9 (Howard et al., 1998)

2.3 Analytical methods

Six general analytical methods have been reported: thin-layer
chromatographic (TLC), liquid chromatographic (LC), mass
spectrometric (MS), post-hydrolysis gas chromatographic, immunochemical and electrophoretic methods (Sydenham & Shephard, 1996;
Shephard, 1998). The majority of studies have been performed using
LC analysis of a fluorescent derivative.

2.3.1 Sampling and preparation procedures

In raw maize, FB_1 is present in both visibly damaged and
undamaged kernels (Bullerman & Tsai, 1994). This means that the
problem that occurs with the mycotoxin aflatoxin, i.e., a few highly
contaminated kernels in otherwise aflatoxin-free kernels, is probably
less of an issue. However, it has been shown that higher levels of
fumonisins are concentrated in visibly damaged kernels (Pascale et al.,
1997). Studies to determine the minimum representative sample in a
lot of maize have not been reported. However, homogeneous material
(CV < 10%) for fumonisin analysis was obtained by grinding
contaminated maize to a particle size less than 2 mm with test portion
sizes of 25 and 10 g (Visconti & Boenke, 1995).

2.3.2 Extraction

Methanol-water (3:1) is the solvent of choice (e.g., Shephard et al.,
1990; Stack & Eppley, 1992; Doko & Visconti, 1994; Scott &
Lawrence, 1994) with a long shaking time or homogenization with a
blender (Sydenham et al., 1992; Bennett & Richard, 1994; Visconti &
Boenke, 1995; Visconti et al., 1995). The use of acetonitrile-water has
also been reported, with conflicting data on its performance relative to
methanol-water (Sydenham et al., 1992a; Bennett & Richard, 1994;
Visconti & Boenke, 1995). Use of an acidic extraction procedure may

lead to higher extraction efficiencies (Zoller et al., 1994; Meister, 1998). However, remarkable variability in extraction efficiency has been reported by several authors, and more work needs to be done to establish the best extraction solvents for various food products.

Clean-up involves the use of solid-phase extraction with strong anion exchange (Shephard et al., 1990) or C_{18} reversed-phase (Ross et al., 1990) or a combination of both (Miller et al., 1993). Improved recoveries can be achieved by using anion exchange instead of reversed-phase material for sample clean-up (Stockenström et al., 1994; Dawlatana et al., 1995). Immunoaffinity columns (Scott & Trucksess, 1997) have also been shown to be useful for clean-up of crude extracts of maize (Ware et al., 1994; Duncan et al., 1998), sweet corn (Trucksess et al., 1995), beer (Scott & Lawrence, 1995) and milk (Scott et al., 1994).

Fumonisins are relatively stable compounds (Alberts et al., 1990; Dupuy et al., 1993a; Le Bars et al., 1994; Visconti et al., 1994; Pascale et al., 1995; Jackson et al., 1996a,b, 1997). A number of factors make them difficult to extract from processed food (Scott, 1993; Bullerman & Tsai, 1994). Binding of FB_1 to maize bran flour occurs at room temperature and above (Scott & Lawrence, 1994). Added iron may also affect recoveries of fumonisin (Scott & Lawrence, 1994). Unknown processing factors or ingredients can change the recovery of fumonisin from cereal products (Scott & Lawrence, 1994). Only 45% of FB_1 present in spiked corn meal was recovered following baking at 175–200 °C for 20 min (Jackson et al., 1997). Fumonisins have been shown to react with reducing sugars at elevated temperatures (Murphy et al., 1996; Lu et al., 1997). The product of the reaction of FB_1 with reducing sugars was identified as N-carboxymethyl-FB_1 (Howard et al., 1998). This product was found in raw corn samples at 4% of the FB_1 levels (Howard et al., 1998). Ammoniation and treatment with base reduces apparent fumonisin concentrations while increasing the concentration of hydrolysed fumonisins without eliminating the toxicity of the treated product, again suggesting analytical difficulties (Norred et al., 1991; Hendrich et al., 1993).

Methods have been reported for the extraction of FB_1 and FB_2 in plasma and urine (Shephard et al., 1992c, 1995c; Shetty & Bhat, 1998), bile of rats and vervet monkeys (Shephard et al., 1994c, 1995a), faeces of vervet monkeys (Shephard et al., 1994b), liver, kidney and muscle of beef cattle (Smith & Thakur, 1996), and milk (Maragos & Richard, 1994; Scott et al., 1994; Prelusky et al., 1996a).

2.3.3 Analysis

Normal phase silica TLC can be used for analysis, with fumonisins being visualized by spraying with *p*-anisaldehyde (Plattner et al., 1990; Sydenham et al., 1990a; Dupuy et al., 1993b). For C_{18} HPLC or TLC, visualization has been accomplished with fluorescamine (Rottinghaus et al., 1992; Miller et al., 1995) and vanillin (Pittet et al., 1992). The detection limit for fumonisins in maize by these methods is 1 mg/kg (Miller et al., 1995). Improved TLC methods with adequate sensitivity are needed, particularly to control maize contamination in developing countries.

A number of fluorescent derivatives have been used for HPLC detection including fluorescamine (Ross et al., 1991a,b), naphthalene-2,3-dicarboxaldehyde/potassium cyanide (Ware et al., 1993; Bennett & Richard, 1994; Scott & Lawrence, 1994), 4-fluoro-7-nitrobenzo-2-oxa-1,3-diazole (Scott & Lawrence, 1992, 1994), 6-aminoquinolyl *N*-hydroxysuccinimidylcarbamate (Velázquez et al., 1995), 9-fluorenylmethyl chloroformate (Holcomb et al., 1993) and *o*-phthaldialdehyde (OPA) (Shephard et al., 1990; Sydenham et al., 1992). In most laboratories, these methods have reported limits of detection or limits of quantification ranging from 5 to 100 μg/kg. The OPA method is widely used and methodology using this derivative has been the subject of international collaborative trials (Thiel et al., 1993; Visconti et al., 1993; Sydenham et al., 1996). Particularly satisfactory results were achieved in the trial by Sydenham et al. (1996) with FB_1 concentrations ranging from 0.5 to 8.0 mg/kg. Relative standard deviations for within-laboratory repeatability ranged from 5.8% to 13.2% for FB_1. Relative standard deviations for between-laboratory reproducibility were 13.9% to 22.2% for FB_1. HORRAT ratios for 7 samples in the test varied from 0.75 to 1.73 for FB_1 (Sydenham et al., 1996). Ratios of less than 2 are considered acceptable. This method has been adopted by the Association of Official Analytical Chemists International as an official method for the analysis of maize.

There are no standardized methodologies for fumonisin analysis in different food products. A method for the extraction and analysis of FB_1 in beer has been reported (Scott & Lawrence, 1994; Scott et al., 1997).

Hydrolysis of samples to the aminopentol chain followed by the GC analysis of the trimethylsilyl or trifluoroacetate derivative by flame ionization detection or mass spectrometry has been reported (Plattner et al., 1990, 1992; Plattner & Branham, 1994). Determination of hydrolysed FB_1 in alkali-processed corn foods by HPLC with fluorescent derivatives has also been reported (Scott & Lawrence, 1996).

Analyses of maize extracts with antibodies reactive with FB_1 (and FB_2 plus FB_3) by direct and indirect assays have been reported (Azcona-Olivera et al., 1992a,b; Usleber et al., 1994; Scott & Trucksess, 1997; Mullett et al., 1998). Detection limits using these methods have been reported to be 0.1-100 µg/litre. In one study, an ELISA method gave higher estimates of fumonisin concentrations compared to GC-MS and HPLC (Pestka et al., 1994).

To a very limited extent, fumonisins have also been determined by capillary zone electrophoresis (CZE). In order to achieve resolution of the FB_1 and FB_2 analogues, samples were derivatized with either 9-fluorenylmethyl chloroformate (Holcomb & Thompson, 1996) or fluorescein isothiocyanate (Maragos, 1995) prior to separation.

As an analytical tool for the determination of fumonisins, MS was initially used as a detector after gas chromatographic separation of the hydrolysed fumonisins (Plattner et al., 1990). Although MS methods using fast-atom bombardment (Plattner & Branham, 1994) and particle beam interfaces (Young & Lafontaine, 1993) have been described, the application of the electrospray interface has led to the greatest advance in the use of MS for fumonisin determination. These methods rely on the LD separation of the underivatized fumonisins and detection of the different analogues as their protonated molecular ions (Doerge et al., 1994; Plattner, 1995; Lukacs et al., 1996; Churchwell et al., 1997). A combined on-line immunoaffinity capture, HPLC/MS method has also been described, and this permits analysis of non-derivatized fumonisins at sub µg/kg levels (Newkirk et al., 1998).

3. SOURCES OF HUMAN EXPOSURE

FB_1 was isolated in 1988 by Gelderblom et al. (1988). It was chemically characterized by Bezuidenhout et al. (1988), and shortly thereafter as "macrofusine" by Laurent et al. (1989a), from cultures of *Fusarium verticillioides* (Sacc.) Nirenberg (*Fusarium moniliforme* Sheldon). A selection of FB_1 occurrence data in maize and food products is given in Table 1 and Appendix 2. A worldwide survey of fumonisin contamination of maize and maize-based products was reported by Shephard et al. (1996a).

FB_1 is produced by isolates of *Fusarium verticillioides, F. proliferatum, F. anthophilum, F. beomiforme, F. dlamini, F. globosum, F. napiforme, F. nygamai, F. oxysporum, F. polyphialidicum, F. subglutinans* and *F. thapsinum* isolated from Africa, the Americas, Oceania, Asia and Europe (Gelderblom et al., 1988; Ross et al., 1990; Thiel et al., 1991a; Nelson et al., 1991, 1992; Chelkowski & Lew, 1992; Leslie et al., 1992, 1996; Rapior et al., 1993; Miller et al., 1993, 1995; Visconti & Doko, 1994; Desjardins et al., 1994; Abbas et al., 1995; Abbas & Ocamb, 1995; Logrieco et al., 1995; Klittich et al., 1997; Musser & Plattner, 1997; Sydenham et al., 1997). A species of *Alternaria* (*A. alternata* f. sp. *lycopersici*) has also been demonstrated to synthesize B fumonisins (Abbas & Riley, 1996). Fumonisins can be produced by culturing strains of the *Fusarium* species that produce these toxins on sterilized maize (Cawood et al., 1991), and yields of up to 17.9 g/kg have been obtained with *F. verticillioides* strain MRC 826 (Alberts et al., 1990). Yields of 500–700 mg/litre for FB_1 plus FB_2 have been obtained in liquid fermentations and high recoveries of the toxins are possible (Miller et al., 1994). The most predominant toxin produced is FB_1. FB_1 frequently occurs together with FB_2, which may comprise 15–35% of FB_1 (IARC, 1993; Diaz & Boermans, 1994; Visconti & Doko, 1994).

Fusarium verticillioides and *F. proliferatum* are amongst the most common fungi associated with maize. These fungi can be recovered from most maize kernels including those that appear healthy (Hesseltine et al., 1981; Bacon & Williamson, 1992; Pitt el al., 1993; Sanchis et al., 1995). The formation of fumonisins in maize in the field is positively correlated with the occurrence of these two fungal species, which are predominant during the late maturity stage (Chulze et al., 1996). These species can cause Fusarium kernel rot of maize, which is one of the most important ear diseases in hot maize-growing areas (King & Scott, 1981; Ochor et al., 1987; De León & Pandey, 1989) and is associated with warm, dry years and/or insect damage (Shurtleff, 1980).

There is a strong relationship between insect damage and Fusarium kernel rot. A field survey demonstrated that the incidence of the European corn borer increased *F. verticillioides* disease and fumonisin concentrations (Lew et al., 1991). Disease incidence was also shown to correlate to populations of thrips (*Frankliniella occidentalis*) (Farrar & Davis, 1991). Hybrids with a thin kernel pericarp were more susceptible to insect wounds, which allowed easier access to the fungus (Hoenisch & Davis, 1994). Hybrids with an increased propensity for kernel splitting had more disease (Odvody et al., 1990). Kernel splitting is worse under drought conditions. Ears infected by *F. graminearum* may be predisposed to *F. verticillioides* infection and fumonisin accumulation (Schaafsma et al., 1993). In maize ears inoculated one week after silk emergence with *F. verticillipodes* fumonisins accumulated in the visibly damaged (mouldy) kernels (Pascale et al., 1997; Desjardins et al. 1998). Sydenham et al. (1995) showed that in lightly contaminated kernels FB_1 was concentrated in the pericarp of the maize kernel.

A study of fumonisin occurrence in hybrids grown across the USA maize belt indicated that hybrids grown outside their range of adaptation had higher fumonisin concentrations (Shelby et al., 1994b), again suggesting the important role of temperature stress. Data from samples collected in Africa, Italy and Croatia also indicate fumonisin accumulation in lines grown outside their area of adaptation (Doko et al., 1995; Visconti, 1996). The occurrence of fumonisin in Ontario, Canada (a cool maize-growing region) was limited to drought-stressed fields (Miller et al., 1995).

Significant fumonisin accumulation in maize occurs when weather conditions favour Fusarium kernel rot, and the severity of ear infection has been found to be a good indicator of fumonisin accumulation in maize ears artificially inoculated with *F. verticillioides* (Pascale et al., 1997). Since monitoring began in the USA, warm, dry years have

Table 1[a] Worldwide occurrence of fumonisin B_1 (FB_1) in maize-based products

Product	Countries	Detected / total	FB_1 (mg/kg)
North America			
Maize	Canada, USA	324/729	0.08–37.9
Maize flour, grits	Canada, USA	73/87	0.05–6.32
Miscellaneous maize foods[b]	USA	66/162	0.004–1.21
Maize feed	USA	586/684	0.1–330
Latin America			
Maize	Argentina, Uruguay, Brazil	126/138	0.17–27.05
Maize flour, alkali-treated kernels, polenta	Peru, Venezuela, Uruguay	5/17	0.07–0.66
Miscellaneous maize foods[b]	Uruguay, Texas-Mexico border	63/77	0.15–0.31
Maize feed	Brazil, Uruguay	33/34	0.2–38.5
Europe			
Maize	Austria, Croatia, Germany, Hungary, Italy, Poland, Portugal, Romania, Spain, United Kingdom	248/714	0.007–250
Maize flour, maize grits, polenta, semolina	Austria, Bulgaria, Czech Republic, France, Germany, Italy, Netherlands, Spain, Switzerland, United Kingdom	181/258	0.008–16
Miscellaneous maize foods[b]	Czech Republic, France, Germany, Italy, Netherlands, Spain, Sweden, Switzerland, United Kingdom	167/437	0.008–6.10
Imported maize, grits and flour	Germany, Netherlands, Switzerland	143/165	0.01–3.35
Maize feed	France, Italy, Spain, Switzerland, United Kingdom	271/344	0.02–70

Table 1 (contd).

Africa			
Maize	Benin, Kenya, Malawi, Mozambique, South Africa, Tanzania, Uganda, Zambia, Zimbabwe	199/260	0.02–117.5
Maize flour, grits	Botswana, Egypt, Kenya, South Africa, Zambia, Zimbabwe	73/90	0.05–3.63
Miscellaneous maize foods[b]	Botswana, South Africa	8/17	0.03–0.35
Maize feed	South Africa	16/16	0.47–8.85
Asia			
Maize	China, Indonesia, Nepal, Philippines, Thailand, Vietnam	361/614	0.01–155
Maize flour, grits, gluten	China, India, Japan, Thailand, Vietnam	44/53	0.06–2.60
Miscellaneous maize foods[b]	Japan, Taiwan	52/199	0.07–2.39
Maize feed	Korea, Thailand	10/34	0.05–1.59
Oceania			
Maize	Australia	67/70	0.3–40.6
Maize flour	New Zealand	0/12	–

[a] This table is a summary of the information in Appendix 2

[b] Includes maize snacks, canned maize, frozen maize, extruded maize, bread, maize-extruded bread, biscuits, cereals, chips, flakes, pastes, starch, sweet maize, infant foods, gruel, purée, noodles, popcorn, porridge, tortillas, tortilla chips, masas, popped maize, soup, taco, tostada

17

greater concentrations than cooler years (Murphy et al., 1993). The direct influence of low moisture and dry weather on fumonisin accumulation could not be proven (Murphy et al., 1996; Pascale et al., 1997), although maize grown under normal conditions in cooler maize-growing areas is not significantly contaminated by fumonisin (Doko et al., 1995; Miller et al., 1995).

Dry milling of maize results in the distribution of fumonisin into the bran, germ and flour (Bullerman & Tsai, 1994). Fumonisin may be present in beer where maize has been used as a wort additive (Scott et al., 1995). Little degradation of fumonisin occurs during fermentation and the fumonisins are found in the spent grain. No toxins can be detected in the distilled ethanol (Bothast et al., 1992; Scott et al., 1995; Bennett & Richard, 1996). Fumonisin is stable in polenta (Pascale et al., 1995), whereas it is hydrolysed, and the pericarp is removed, by nixtamalization, i.e. the treatment of maize-based foods with calcium hydroxide and heat (Hendrich et al., 1993). FB_1 has been shown to form N-(carboxymethyl)-FB_1 when heated in the presence of reducing sugars (Howard et al. 1998), and the latter substance has been detected in raw corn (Howard et al., 1998).

FB_1 is not significantly transferred into pork, chicken meat or eggs (Prelusky et al., 1994, 1996a; Vudathala et al., 1994), but a small amount accumulates in the liver and kidney of pigs as a function of exposure (Prelusky et al., 1996b; see also section 6.2). Fumonisin is not significantly transferred into milk from short-term dietary exposure (Scott et al., 1994; Prelusky et al., 1996a), and FB_1 was found in only one of 165 samples of milk from Wisconsin, USA at a level close to 5 ng/ml (Maragos & Richard, 1994).

4. ENVIRONMENTAL TRANSPORT, DISTRIBUTION AND TRANSFORMATION

Maize is the only commodity that contains significant amounts of fumonisins. It is consumed either directly or processed into products for human or animal consumption. Because fumonisins are known to be heat stable (Dupuy et al., 1993a; Howard et al., 1998), light stable (IARC, 1993), water soluble (US NTP, 1999), poorly absorbed, poorly metabolized and rapidly excreted by animals (see sections 6.1 to 6.5), most fumonisin will eventually end up being recycled into the environment in a manner that will concentrate its spatial distribution. The amount that enters the environment may be quite large. For example, in the USA, maize production exceeds 200 million tonnes per year. The concentration of FB_1 and FB_2 in field maize in the USA often exceeds 1 g/tonne of maize (Murphy et al., 1993 and Appendix 2). There is some evidence that fumonisins can be metabolized by some microorganisms (Duvick et al., 1994, 1998). However, little is known about the environmental fate of fumonisin after it is either excreted or processed.

5. ENVIRONMENTAL LEVELS AND HUMAN EXPOSURE

Table 1 summarizes the results of a number of surveys on the natural occurrence of FB_1 in maize and maize-based foods and feeds (see Appendix 2 for more detail). The list is not exhaustive of the surveys carried out worldwide as there is continual production of similar data from every corner of the globe. Based on Table 1, 60% of the 5211 samples analysed have been found to be contaminated with FB_1, the highest incidences of contamination being in Oceania (82% of 82 samples) and Africa (77% of 383 samples), followed by Latin America (85% of 266 samples), North America (63% of 1662 samples), Europe (53% of 1918 samples) and Asia (52% of 900 samples).

The data show that levels and incidence of contamination vary considerably in relation to the commodities tested and the source. The highest incidence was recorded in maize feeds (82% of 1112 samples), followed by ground maize products, such as flour, grits, polenta, semolina and gluten (73% of 517 samples), maize kernels (52% of 2525 samples) and miscellaneous maize foods (40% of 892 samples).

FB_1 levels in animal feedstuffs can be exceptionally high, and reached maximum values of 330, 70, 38, 9 and 2 mg/kg in North America (USA), Europe (Italy), Latin America (Brazil), Africa (South Africa) and Asia (Thailand), respectively. The majority of the highly contaminated feeds were implicated in cases of equine leukoencephalomalacia, porcine pulmonary oedema and other mycotoxicoses.

In maize kernels available commercially or from experimental or breeding stations, FB_1 has been detected in 96% (of 70 samples), 91% (of 138 samples), 76% (of 260 samples), 59% (of 614 samples), 44% (of 729 samples) and 35% (of 714 samples) of samples from Oceania, Latin America, Africa, Asia, North America and Europe, respectively. Maximum FB_1 levels were 40.6 mg/kg (Australia), 27 mg/kg (Argentina), 117 mg/kg (South Africa), 155 mg/kg (China), 38 mg/kg (USA) and 250 mg/kg (Italy).

The list of commercial retail foods subject to fumonisin contamination (Table 1) includes maize flour, grits, polenta, semolina, maize snacks, cornflakes, sweet maize, canned maize, frozen maize, extruded maize, bread, maize-extruded bread, biscuits, cereals, chips, pastes, starch, infant foods, gruel, purée, noodles, popcorn, porridge, tortillas, tortilla chips, masas, popped maize, soup, taco and tostada.

Of these samples, the global incidence of contamination in non-treated or minimally treated maize products (flour, grits, polenta, semolina) was 73% out of 517 samples analysed. The highest FB_1 levels were recorded in Europe (16 mg/kg), followed by North America (6.3 mg/kg), Africa (3.6 mg/kg), Asia (2.6 mg/kg) and Latin America (0.7 mg/kg). In the remaining food products (892 samples) the incidence of contamination was 40%, the highest level (6.1 mg/kg FB_1) being found in a sample of extruded maize from Italy. Generally processed maize foods have lower levels and incidence of contamination than non-treated maize. These differences might be the results of dilution of maize in food commodities, or may depend on the differences in maize cultivar or quality requirements for various destinations.

Apart from maize and maize products, fumonisins have seldom been found in other food products, such as rice (Abbas et al., 1998), asparagus (Logrieco et al., 1998), beer (Torres et al., 1998) and sorghum (Shetty & Bhat, 1997). Surveys on other cereals, such as wheat, rye, barley and oats, did not show the occurrence of the toxin (Meister et al., 1996).

Human exposure estimates have been made for fumonisins in several countries, including Switzerland, Canada, South Africa, USA and the Netherlands (Zoller et al., 1994; Contaminants Standards Monitoring and Programs Branch, 1996a,b; Gelderblom et al., 1996b; Kuiper-Goodman et al., 1996; Humphreys et al., 1997; Marasas, 1997; de Nijs, 1998). Human exposure estimates of 0.017–0.089 µg/kg body weight per day have been prepared for Canada for the period 1991 to early 1995 (Kuiper-Goodman et al., 1996). For the USA, a preliminary estimate of human exposure to fumonisins for maize eaters was 0.08 µg/kg body weight per day (Humphreys et al., 1997). The mean daily intake of fumonisins in Switzerland is estimated to be 0.030 µg/kg body weight per day (Zoller et al., 1994).

Based on the daily average intakes of maize and maize products of 3 g (general population average), 42 g (regular maize product eaters) and 162 g (individuals with gluten intolerance) in the Netherlands, the respective population groups had an estimated daily intake of 4, 57 and 220 µg FB_1 per person, respectively, based on a mean FB_1 content of 1.36 mg/kg maize produce. De Nijs et al. (1998a) estimated conservatively that 97% of individuals with gluten intolerance had a daily exposure of at least 1 µg FB_1 and 37% at least 100 µg, while the proportions of the general population exposed to these levels of FB_1 were 49% and 1%, respectively (de Nijs, 1998; de Nijs et al., 1998a).

Thiel et al. (1992) estimated that human exposures in the Transkei, South Africa, are 14 and 440 µg FB_1/kg body weight per day for healthy and mouldy corn, respectively. More recent estimates of the probable daily intake (PDI) of South Africans are summarized in Table 2. These vary from 1.2 to 355 µg/kg body weight per day in rural blacks in Transkei consuming home-grown mouldy maize (Gelderblom et al., 1996b; Marasas, 1997).

These exposure estimates will vary considerably according to the source and extent of maize in the diet as well as the extent of Fusarium kernel rot prevalent in the harvested crop.

Occupational inhalation exposure could be a problem. In addition to the presence of fumonisins in maize dust, FB_1 is present in the spores and mycelia of *F. verticillioides* (Tejada-Simon et al., 1995). No data have been collected on airborne levels of fumonisin during the harvesting, processing and handling of fumonisin-contaminated maize.

Table 2. Probable daily intake of fumonisin in South Africa[a]

Product	Country of origin	No. of samples	Mean FB$_1$ + FB$_2$ concentration (μg/kg)	Probable daily intake (μg/kg body weight per day)	
				Rural population	Urban population
Commercial maize	South Africa	68	400	2.6	1.6
Commercial maize	South Africa	209	300	2.0	1.2
Corn meal	South Africa	52	200	1.3	0.8
Home-grown maize	South Africa[b]	18	7100	46.6	28.0
Home-grown maize	South Africa[c]	18	54 000	354.9	212.9
Imported maize	USA[d]	1682	1100	7.2	4.3
Maize consumption (g/70 kg body weight per day)				460	276

[a] From: Marasas (1997)
[b] Transkei, from individual farms in high oesophageal cancer area, healthy maize
[c] Transkei, from individual farms in high oesophageal cancer area, mouldy maize
[d] Imported in 1993

6. KINETICS AND METABOLISM IN ANIMALS

There have been no reports on the kinetics and metabolism of fumonisins in humans. Because fumonisins are known to be consumed by farm animals and are the causative agent or a suspected contributing factor in farm animal diseases, an effort has been made to understand the kinetics and metabolism in cows, pigs and poultry. Thus, this chapter will summarize results of studies on both laboratory and farm animals.

To date, published studies with radiolabelled FB_1 or FB_2 have been conducted with either [21,22-^{14}C]fumonisins, biosynthesized using L-[methyl-^{14}C]methionine (Plattner & Shackelford, 1992; Alberts et al., 1993), or [U-^{14}C]FB_1 labelled using [1,2-^{14}C]acetate (Blackwell et al., 1994). In these studies the final [^{14}C]fumonisins had a specific activity of < 1 mCi/mmol and radiochemical purity of > 95%. Several studies have used unlabelled fumonisins with reported purities ranging from 70% (Hopmans et al., 1997) to 98% (Prelusky et al., 1996a).

Briefly, FB_1 is: poorly absorbed when dosed orally; it is rapidly eliminated from plasma or circulation and recovered in faeces; biliary excretion is important; enterohepatic cycling is clearly important in some animals; small amounts are excreted in urine; a small but persistent (and biologically active) pool of [^{14}C]label appears to be retained in liver and kidney; and some is degraded to partially hydrolysed FB_1 in the gut of vervet monkeys. In a study with FB_2 in rats, the results were similar to those of FB_1 (Shephard et al., 1995b).

6.1 Absorption

There are no reports available of fumonisin absorption through inhalation or dermal exposure. However, because fumonisins are present in *F. verticillioides* cells (mycelia, spores and conidiophores) (Tejada-Simon et al., 1995), there is a potential for absorption through inhalation or buccal exposure. The risk from absorption due to dermal exposure would seem slight, since fumonisins are very water soluble and, typically, polar compounds do not easily penetrate the undamaged skin (Flynn, 1985).

The quantity of FB_1 detected in plasma after oral dosing in pigs, laying hens, vervet monkeys, dairy cows and rats is very low. In rats (BD IX, Sprague-Dawley or Wistar) administered $[^{14}C]FB_1$ orally, accumulation of ^{14}C-labelled compounds in tissues is also very low, suggesting that absorption is very poor (negligible to < 4% of dose) (Shephard et al., 1992a,b, 1994c; Norred et al., 1993). Similar results indicating that fumonisins are poorly absorbed (2 to < 6% of dose) have been reported in vervet monkeys, dairy cows and pigs (Prelusky et al., 1994, 1995, 1996a,b; Shephard et al., 1994a,b). In orally dosed laying hens and dairy cows, systemic absorption based on plasma levels and accumulation of ^{14}C-labelled compounds in tissues has been estimated to be less than 1% of dose (Scott et al., 1994; Vudathala et al., 1994; Prelusky et al., 1996a). A study using beef cattle fed *F. verticillioides* culture material (corn grits) containing FB_1 plus FB_2 (530 mg/kg) found that the majority of the fumonisin dose was recovered unmetabolized in faeces, and only traces were detected in blood and urine (Smith & Thakur, 1996). Following single gavage doses of 1 or 5 mg/kg body weight to cows, no FB_1 or known metabolites could be found in the plasma, indicating no or very limited bioavailability in ruminants (Prelusky et al., 1995). Rumen metabolism may reduce the bioavailability of FB_1 as the hydrolysed form of FB_1 comprised 60-90% of the total amount of FB_1 found in faeces. In non-ruminants the parent compound was the dominant species present (Rice & Ross, 1994).

6.2 Distribution

In rats and pigs orally dosed with $[^{14}C]FB_1$, the ^{14}C label is distributed to most tissues, with the liver and kidney containing the highest concentration of radiolabel (Shephard et. al., 1992b; Norred et al., 1993; Prelusky et al., 1994, 1996a,b; Haschek et al., 1996). Typically, the liver contains more ^{14}C label than the kidney, although in the study by Norred et al. (1993) the measured radioactivity in the kidney was greater than in the liver. In chickens and dairy cows the poor absorption of $[^{14}C]FB_1$ (< 1% of oral dose) was reflected in the fact that only trace amounts of radioactivity were recovered in tissues (Prelusky et al., 1996a), no residues were recovered in eggs of laying hens (Vudathala et al., 1994) and no FB_1 or aminopentol hydrolysis products were recovered in milk (Scott et al., 1994; Prelusky et al., 1996a). In pregnant rats dosed intravenously with $[^{14}C]$fumonisin, approximately 14.5% and 4% of the dose were recovered in the liver and kidney, respectively, after 1 h (Voss et al., 1996a). Based on the known pharmacokinetics (Norred et al., 1993) in the rat, 1-h exposure and intravenous injection were chosen so as to optimize the presentation in blood of the $[^{14}C]FB_1$ to the placentae. In contrast to liver and kidney, the uteri contained 0.24 to 0.44%, individual placentae contained 0 to 0.04%, and total fetal recovery was ≤ 0.015% of dose/dam (Voss et al., 1996a). Recent studies have confirmed the lack of placental transfer of FB_1 in rats (Collins et al., 1998a,b) and rabbits (LaBorde et al., 1997).

FB_1 inhibition of the enzyme sphinganine N-acyltransferase results in a large increase in intercellular free sphinganine (Wang et al., 1991; Yoo et al., 1992). In animal tissues the fumonisin-induced increase in free sphinganine tends to parallel the distribution of ^{14}C label reported in the studies cited above using $[^{14}C]FB_1$. For example, relative to other tissues examined, liver and kidney in rabbits, pigs and catfish showed the greatest increases in free sphinganine following exposure of animals to fumonisins or consumption of diets containing fumonisins (Goel et al., 1994; Gumprecht et al., 1995). The free sphinganine concentration in tissues has been shown to be an easily detectable biomarker for exposure to fumonisins (Riley et al., 1994c), although it has not been validated as a biomarker in humans.

6.3 Elimination, excretion and metabolic transformation

When $[^{14}C]FB_1$ is dosed by intraperitoneal or intravenous injection in rats (BD IX, Sprague-Dawley or Wistar), initial elimination (subsequent to the distribution phase) is rapid (half-life of approximately 10–20 min) with little evidence of metabolism (Shephard et al., 1992a,b, 1994c; Norred et al., 1993). In rats the elimination kinetics based on intraperitoneal or intravenous dosing are consistent with a one- (Shephard et al., 1992b) or two-compartment model (Norred et al., 1993). Because FB_1 is poorly absorbed from the rat gastrointestinal tract and extensively distributed in rat tissues (Norred et al., 1993), the tissue elimination kinetics following oral dosing is not as easily described. In vervet monkeys, as in rats, the ^{14}C label is widely distributed and rapidly eliminated (half-life of 40 min) after intravenous injection (Shephard et al., 1994a,b). The elimination kinetics following oral dosing in a non-human primate has not been determined. Following single intravenous injection of 0.05 or 0.20 mg FB_1/kg body weight to cows, the toxin is cleared rapidly from the blood. A two-compartment model (half-lives of < 2 and 15–18 min, respectively) satisfactorily described the plasma kinetics. No toxin could be detected 120 min after dosing. No known metabolites were detected in the plasma (Prelusky et al., 1995).

In pigs, clearance of $[^{14}C]FB_1$ from blood following an intravenous injection was best described by a 3-compartment model (half-lives of 2.5, 10.5 and 183 min, respectively), and cannulation of the bile duct (bile removed) resulted in a much more rapid clearance (best described by a 2-compartment model). The effect of bile removal was observed whether dosing was intravenous or intragastric (Prelusky et al., 1994, 1996a). The half-life in pigs dosed intragastrically without bile removal was determined to be 96 min (Prelusky et al., 1996a). The studies with pigs strongly support the importance of enterohepatic circulation of FB_1 in pigs. As in the study with rats, the majority of ^{14}C label dosed orally was recovered in faeces (approximately 90%) with less than 1% recovered in urine (Prelusky et al., 1994, 1996a). In the LLC-PK$_1$ renal cell line, uptake of $[^{14}C]FB_1$ reached an equilibrium concentration with the extracellular $[^{14}C]FB_1$ concentration after 4 to 16 h, and kinetics were indicative of a simple diffusion process (Riley & Yoo, 1995). Efflux was rapid with a half-life of less than 5 min.

Following intravenous injection into rats, FB_1 is excreted unchanged in bile (Norred et al., 1993; Shephard et al., 1994c). In vervet monkeys there is evidence of metabolism to partially hydrolysed (one propane tricarboxylic acid residue removed) FB_1, and to a much lesser extent the fully hydrolysed (both propane tricarboxylic acid residues removed) aminopentol backbone, in faeces while in urine 96% of the ^{14}C label was recovered as FB_1 (Shephard et al., 1994a,b). Metabolism was most likely occurring in the gut since partially hydrolysed and fully hydrolysed FB_1 were recovered in the faeces but not in the bile of vervet monkeys (Shephard et al., 1995a). Because hydrolysed FB_1 and FB_1-fructose adduct can be formed during processing, Hopmans et al. (1997) evaluated the excretion of these products and FB_1 in Fischer-344 rats. Based on the amount of each FB_1-related compound recovered in urine and faeces, it was concluded that hydrolysed FB_1 and the FB_1-fructose adduct were better absorbed than FB_1 (Hopmans et al., 1997).

Dairy cows dosed with pure FB_1 either orally (1.0 and 5.0 mg FB_1/kg body weight) or by intravenous injection (0.05 and 0.20 mg FB_1/kg body weight) showed no detectable residues of FB_1, AP_1 (the aminopentol hydrolysis product of FB_1) or their conjugates in the milk (Scott et al. 1994). FB_1 does not react with monoamine or diamine oxidase (Murphy et al., 1996). *In vitro* studies using rat primary hepatocytes and microsomal preparations (Cawood et al., 1994) or studies with the $LLC-PK_1$ renal epithelial cell line (Riley & Yoo, 1995) indicated that there was no metabolism of FB_1 in these systems.

Repeated intraperitoneal injection of FB_1 resulted in induction of cytochrome P-4501A1 and P-4504A1 activities (Martinez-Larrañaga et al., 1996). However there is no evidence that fumonisin is metabolized by P-450 enzymes. Whether or not the induction was due to a direct interaction between fumonisins and the metabolizing systems could not be determined. However, it has been shown that some of the same sphingolipid metabolites that are altered in fumonisin-treated animals also mediate the cytokine-induced alterations in P-4502C11 in rat hepatocytes (Nikolova-Karakashian et al., 1997).

6.4 Retention and turnover

$[^{14}C]FB_1$ is widely distributed in tissues of the rat and pig. However, only the liver and kidney retain small but persistent amounts of ^{14}C label based on measured radioactivity (Norred et al., 1993; Prelusky et al., 1994, 1996b). In rats given three repeated oral doses, once accumulated, the measured radioactivity in liver and kidney remained unchanged for at least 72 h after the last intragastric dose (Norred et al., 1993). In pigs, it was estimated that exposure to dietary FB_1 at 2–3 mg/kg in feed would require a withdrawal period of at least 2 weeks for the ^{14}C label to be eliminated from the liver and kidney (Prelusky et al., 1996b). The chemical nature of the ^{14}C-labelled material retained in liver and kidney was primarily FB_1.

In vitro studies with rat primary hepatocytes and the cultured kidney cell line $LLC-PK_1$ also indicate that a low but persistent pool of ^{14}C-labelled material is retained inside cells long after the rapidly diffusible pool of $[^{14}C]$fumonisin has exited the cells (Cawood et al., 1994; Riley et al., 1998). This retained pool appears to be capable of maintaining the elevation of cellular (LLC-PK$_1$ cells) and urinary (in rats) free sphingoid base concentration, a biomarker of fumonisin exposure (Solfrizzo et al. 1997b; Riley et al., 1998; Wang et al., 1999).

6.5 Reaction with body components

Fumonisins are potent inhibitors of the enzyme sphinganine (sphingosine) *N*-acyltransferase in the *de novo* sphingolipid biosynthesis and sphingolipid turnover pathways (Wang et al., 1991). The consequences of this reaction will be discussed in sections 7.8 and 7.9. FB$_1$ may also interact directly with protein kinase C (Yeung et al., 1996) and/or with mitogen-activated protein kinases (Wattenberg et al., 1996). The only other information concerning reaction with body components is that FB$_1$ does not bind strongly to chicken plasma proteins (Vudathala et al., 1994).

Cytotoxicity studies in primary rat hepatocytes and binding studies using subcellular fractions indicated that ^{14}C-labelled FB$_1$ binds tightly to hepatocytes and microsomal and plasma membrane fractions (Cawood et al. 1994). FB$_1$ has been shown to interact directly with liposomes (Yin et al., 1996). Since fumonisins are water soluble, are not accumulated and are rapidly eliminated, the toxicological significance of this finding is unclear.

7. EFFECTS ON ANIMALS AND *IN VITRO* TEST SYSTEMS

7.1 Laboratory animals and *in vitro* test systems

The studies described below used either purified FB_1, naturally contaminated corn or cultures of *Fusarium*. It is generally accepted that the *in vivo* toxicity of *Fusarium verticillioides* MRC 826 culture material is the result of its high FB_1 content. Culture materials other than MRC 826 may contain several other products such as other fumonisins, fusarins, moniliformin and beauvericin.

7.1.1 Single exposure

In the male Sprague-Dawley rat, intravenous injection of FB_1 (95% purity) at 1.25 mg/kg body weight resulted in renal lesions localized to the tubules in the outer medulla and consisted of both proliferation and death of cells. An increased number of mitotic figures, stained with 5-bromo-2'-deoxyuridine (not quantified), and apoptosis followed by severe nephrosis were observed (Lim et al., 1996). Cell proliferation was also detected in the liver 24 h after dosing, but was not significantly different from control values at later times (day 2 to day 5). In the oesophagus, increased cell proliferation was measured on day 3, but this returned to the control level on day 5. While kidney lesions were reported as severe, the increased mitotic activity in the liver and oesophagus occurred in the absence of morphological injury (Lim et al., 1996).

No information is available on the toxicological effects of single exposure to FB_1 by the inhalation or dermal route.

7.1.2 Repeated exposure

7.1.2.1 Body weight loss

In male BD IX rats consuming a diet containing 1 g FB_1/kg during a 4-week promotion treatment, the mean body weights were 50% lower than those of non-treated rats ($P < 0.0001$), both with and without initiation with diethylnitrosamine (DEN) (Gelderblom et al., 1988). Similarly, the body weight gains of male Fischer rats fed the same concentration of FB_1 over a 26-day initiation period were 80% lower than those of the controls ($P < 0.0025$) (Gelderblom et al., 1992b).

Male Fischer rats fed diets containing 1 g FB_1 (and FB_2 plus FB_3) per kg over a 21-day initiating period started to lose weight within the first week, and the level of the FB compounds had to be reduced by half (Gelderblom et al., 1993). Body weight losses were first observed in rats fed FB_2, where a significant ($P = 0.008$) reduction compared to the controls was recorded after 4–5 days. In the case of FB_1 and FB_3, significant ($P = 0.01$) reductions in body weight occurred after 7–8 days. Body weight loss induced by FB_1 and FB_2 was significantly ($P = 0.001$) higher than that induced by FB_3 (Gelderblom et al., 1993). In female Sprague-Dawley rats administered purified FB_1 at gavage doses of 0, 1, 5, 15, 35 or 75 mg FB_1/kg body weight per day for 11 consecutive days, significant depression of body weight and food consumption was observed at 35 and 75 mg FB_1/kg body weight per day (Bondy et al., 1998).

The reduction in body weight gain of male Fischer rats induced by FB_1 is apparently due to a feed refusal effect (Gelderblom et al., 1994). During a feeding study over 21 days, the body weight gains of rats receiving 750, 500, 250 and 100 mg FB_1/kg diet were significantly ($0.01 < P < 0.05$) lower than those of the controls as well as those of rats receiving 50 and 25 mg/kg. Based on the weekly feed intake profiles, the reduction in body weight gain was accompanied by a concomitant reduction in feed intake. The reduction in feed intake was overcome after the second week, resulting in a feed intake similar to that of the controls at the end of the 21-day initiating treatment (Gelderblom et al., 1994).

In male Fischer-344/N Nctr BR rats, exposure to 234 and 484 mg FB_1/kg diet resulted in 10% and 17%, respectively, less gain in body weight after 28 days of feeding in the range-finding study by the US National Toxicology Program (US NTP, 1999). Female rats had decreased body weight only at 484 mg FB_1/kg diet.

In the NTP 2-year carcinogenicity study (US NTP, 1999) (see section 7.1.6.1), there was no difference in body weight or feed consumption in male or female Fischer-344/N Nctr BR rats or $B6C3F_1$/Nctr BR mice fed FB_1 when compared to rats or mice on control diets.

The characteristic reduction in the body weight of rats induced by FB$_1$, was also induced by FB$_2$, FB$_3$ and the monomethyl esters of FB$_1$ (MME, an artefact of the isolation procedure of FB$_1$ and a minor contaminant of FB$_1$ preparations), and to a much lesser extent by the *N*-acetylated analogue FA$_1$, but not by the aminopolyol hydrolysis products AP$_1$ and AP$_2$ or the tricarbalyllic acid moiety (TCA) (Gelderblom et al., 1993).

7.1.2.2 *Hepatocarcinogenicity and nephrotoxicity*

The acute toxicity of FB$_1$ was tested by dosing four male BD IX rats orally with 240 mg FB$_1$/kg body weight per day (Gelderblom et al., 1988). Three of the four rats died within 3 days and exhibited toxic hepatosis characterized by scattered single-cell necrosis accompanied by mild fatty changes, hydropic (i.e., the abnormal accumulation of serous fluid in the cellular tissue or in a body cavity) degeneration and hyaline droplet degeneration. Hepatocellular nuclei varied in size and some were markedly enlarged. In addition to the hepatotoxic changes, fatty changes and scant necrosis were present in the proximal convoluted tubules of the kidney, prominent lymphoid necrosis was observed in Peyer's patches, and severe disseminated acute myocardial necrosis and severe pulmonary oedema were observed in two of the rats (Gelderblom et al., 1988).

In a separate experiment, male BD IX rats were dosed orally with 48 mg FB$_1$/kg body weight per day for 12 days, followed by 70 mg FB$_1$/kg body weight per day for the remaining 9 days of the experiment (Gelderblom et al., 1988). In the rats killed after 21 days, chronic toxic hepatosis was present and characterized by marked hydropic degeneration, single-cell necrosis and a few hyaline droplets, early bile duct proliferation and fibrosis, and enlargement of hepatocellular nuclei (Gelderblom et al., 1988).

In the livers of rats killed after 33 days on a diet containing 1 g FB$_1$/kg, the hepatic changes were similar to those described above, but more advanced (Gelderblom et al., 1988). The proliferation of bile ducts and fibrosis caused distortion of the lobular structure of the liver and, together with the development of hyperplastic nodules, gave the liver a distinctly nodular appearance. The authors reported that many nuclei were enlarged in hepatic cells and numerous mitotic figures, some of which were abnormal, were present. The lesions in the kidneys were similar, but less severe, than those seen in the rats that died within 3 days (Gelderblom et al., 1988).

In male Fischer rats fed a diet containing 1 g FB_1 (90-95% pure) per kg during an initiating period of 26 days, followed by partial hepatectomy and a promoting regimen of 2-acetyl-aminofluorene (2-AAF) and carbon tetrachloride, early pathological changes in the liver were very similar to those described above (Gelderblom et al., 1992b). Early hepatocyte nodules were evident as discrete focal changes in hepatocytes characterized by somewhat bigger cells that displayed more mitotic figures than the cells in the surrounding liver and also showed vacuolization. Another prominent pathological feature was the mild-to-moderate proliferation of bile ducts (Gelderblom et al., 1992b). Similar hepatic changes have been described in male Fischer rats fed diets containing, at a level of 0.5–1 g/kg, FB_1, FB_2, FB_3 and MME during an initiating period of 21 days followed by a promoting treatment of 2-AAF and partial hepatectomy (Gelderblom et al., 1993). The short-term toxicological effects in rats of FB_2 and FB_3 are similar to those of FB_1 (Gelderblom et al., 1992a).

Changes including hydropic swelling, hyaline droplet accumulation, single-cell necrosis, increased mitotic figures, lipid accumulation, fibrosis, and bile duct proliferation were also observed in the liver of male Fischer rats that died after gavage treatment with 50 mg FB_1/kg body weight in 6 dosages over 11 days (Gelderblom et al., 1994).

A 4-week exposure of Sprague-Dawley rats to aqueous extracts of *Fusarium verticillioides* (MRC 826) cultures (containing fumonisins) resulted in decreased body weights, increased serum alanine and aspartate aminotransferase and alkaline phosphatase activities, decreased relative liver weights and microscopic liver lesions in rats (Voss et al., 1990).

Male and female Sprague-Dawley rats (3 of each sex per group) were fed diets containing 0, 15, 50 and 150 mg/kg of FB_1 ($\geq$ 99% pure) for 4 weeks (Voss et al., 1993). No significant differences in weight gain or food consumption were found, but significant increases in serum triglycerides, cholesterol and alkaline phosphatase confirmed that a dietary level of 150 mg/kg was hepatotoxic to both sexes. Histopathological changes in the liver of these rats were characterized by scattered single-cell hepatocellular necrosis, variability in nuclear size and staining and hepatocellular cytoplasmic vacuolation. Nephrosis, consisting of focal cortical proximal tubular epithelial basophilia, hyperplasia and single cell necrosis or pyknosis, was found in males fed ≥ 15 mg/kg and in females fed ≥ 50 mg/kg (Voss et al., 1993). The incidence and severity of ultrastructural alterations in kidney and liver were closely correlated with increased sphinganine concentration in tissues, serum and urine (Riley et al., 1994a).

The apparent no-observed-effect level (NOEL) for renal toxicity in FB_1-fed rats was less than the NOEL for hepatic effects (4.1 < NOEL < 13.6 mg/kg diet for 28 days), and renal toxicity was more severe in males (NOEL < 1.4 mg/kg diet for 28 days) than females (1.4 < NOEL < 4.1 mg/kg diet for 28 days). Furthermore, liver lesions found in females appeared (subjectively) more advanced than those found in males. The results of subacute toxicity studies (7.5 and 10 mg/kg body weight per day for 4 days) (Bondy et al., 1995; Suzuki et al., 1995) and of an independent (Tolleson et al., 1996a) 4-week study in Fischer-344 rats fed 0, 99, 163, 234 or 484 mg FB_1/kg diet corroborated the findings of nephrotoxicity by Voss et al. (1993). Hepatopathy of the same type was found in males fed > 234 mg/kg diet and females fed ≥ 163 mg/kg diet. Nephropathy was found in males from all FB_1-fed groups and in females fed ≥ 163 mg/kg diet (Tolleson et al., 1996a). Apoptotic hepatocytes and renal proximal tubule epithelial cells were accompanied by cell proliferation in Fischer-344 rats, suggesting that fumonisin induces or accelerates programmed cell death in both liver and kidney (Tolleson et al. 1996a; US NTP, 1999).

In male and female B6C3F₁ mice administered FB_1 at gavage doses ranging from 1 to 75 mg FB_1/kg body weight per day for 14 days, effects on liver, bone marrow, adrenals and kidneys were observed. In general, however, the degree of change observed indicates that mice are not as sensitive to FB_1 toxicity as rats (Bondy et al., 1995, 1997).

In B6C3F₁ mice fed 99 to 484 mg FB_1/kg diet for 4 weeks, the liver, not the kidney, was the target organ (US NTP, 1999). As for rats, the NOEL was lower in females as liver lesions were found in the females of all FB_1-fed groups, while in males hepatopathy was confined to the highest dose group. In male BALB/c mice dosed subcutaneously (0.25 to 6.25 mg FB_1/kg body weight per day), a dose-dependent increase in apoptosis was observed in both liver and kidney (Sharma et al., 1997).

To obtain dose–response data under longer-term exposure conditions, Fischer-344 rats and $B6C3F_1$ mice were fed diets containing 0, 1, 3, 9, 27 or 81 mg FB_1/kg diet for 13 weeks (Voss et al., 1995). In rats, toxicity was confined to the kidneys. Lesions of the proximal tubule located in the outer medulla (sometimes referred to as the corticomedullary junction) were found in males fed ≥ 9 mg/kg diet and in females fed 81 mg/kg diet. Qualitatively these lesions were of the same type as those found in the 4-week study (Voss et al., 1993). No differences in the incidence or severity of nephropathy between rats examined after 4 (n = 5 rats/group) or 13 (n = 10/group) weeks were found.

Renal lesions were accompanied by decreased relative kidney weight (as a percentage of body weight), which was found in males fed ≥ 27 mg/kg diet for 4 weeks and in both sexes fed ≥ 9 mg/kg diet for 13 weeks. Serum creatinine was increased after 13, but not 4, weeks in males fed ≥ 27 mg/kg diet and in females fed 81 mg/kg diet (Voss et al. 1995).

In mice, hepatopathy and serum chemical evidence of liver dysfunction were found after 13 weeks in females fed 81 mg FB_1/kg diet (Voss et al., 1995). Liver lesions in female mice were primarily centrilobular, although some midzonal involvement and apparent "bridging" between adjacent central areas were evident. Single cell hepatocyte necrosis, cytomegaly, increased numbers of mitotic figures, some mixed infiltration of neutrophils and macrophages were present and, in more advanced lesions, the loss of hepatocytes caused an apparent collapse around the central vein. Hepatopathy was not found in male mice and FB_1-related kidney lesions did not occur in either sex. A few macrophages containing minimal to mild amounts of cytoplasmic pigment, presumably ceroid, were also found in the adrenal cortex of high-dose (81 mg/kg diet) females only.

Taken together, the findings from 4-week and 90-day toxicity studies in rats and mice (Voss et al., 1993, 1995; Tolleson et al., 1996a; US NTP, 1999) indicate that the liver is a target organ in both species, and the data seem to indicate that females exhibit hepatic effects at lower doses than males. In rats, however, the kidney is also an important target organ and, in contrast to liver, the males were affected at lower doses.

7.1.2.3 Immunotoxicity

There have been very few studies that address directly the potential for fumonisins to modify immune response *in vivo*. Nonetheless, there are many studies with fumonisins or fumonisin-containing diets that show either altered function of blood cells *in vitro* or changes in haematological parameters *in vivo*. Fumonisins are inhibitors of ceramide synthase (see section 7.3) and ceramide and glycosphingolipids are important signalling molecules and recognition sites in the cellular immune response and attachment sites for many infectious agents and microbial toxins (Ballou et al., 1996; Merrill et al., 1997a).

In a study with pure FB$_1$, changes in selected haematological parameters in pigs were reported at dietary levels as low as 1 mg/kg (Rotter et al., 1996). Consumption of culture-material diets (MRC 826) containing fumonisins decreased the ability to clear *Pseudomonas aeruginosa* and inhibited pulmonary interstitial macrophage function (Smith et al., 1996c). It was hypothesized that pulmonary intravascular macrophage (PIM) dysfunction could contribute to increase susceptibility to microbial diseases (Smith et al., 1996c).

Cytokine production has been shown to be modified by exposure to fumonisin. For example, serum tumour necrosis factor-α (TNF-α)-like activity was increased in pigs fed culture material (M 1325 = MRC 826) containing 150 mg/kg fumonisins (Guzman et al., 1997). Fumonisin-induced changes in the TNF pathway have also been seen in lipopolysaccharide (LPS)-stimulated macrophages collected from BALB/c mice dosed with pure FB$_1$ (Dugyala et al., 1998).

Immunosuppression in chickens was produced in birds fed maize cultured with *F. verticillioides* (MRC 826) (Marijanovic et al., 1991).

Broiler chicks fed diets containing 10 mg pure FB_1/kg diet, or diets formulated from *Fusarium verticillioides* (MRC 826) culture material to contain 30 to 300 mg FB_1/kg diet, had reduced spleen and/or bursa weights and altered haematological parameters (Espada et al., 1994, 1997).

In male and female rats (10 rats/group) gavaged daily for 14 days with doses of 0, 5, 15 or 25 mg FB_1/kg body weight per day, a significant dose-related linear trend toward decreased plaque-forming cell number per 10^6 spleen mononuclear leukocytes (PFC per 10^6 splenocytes) ($P = 0.003$) and PFC per spleen cells ($P = 0.001$) was observed in the male rats. However, the PFC numbers in female rats were not affected significantly by treatment ($P > 0.05$) (Tryphonas et al., 1997).

7.1.3 Skin and eye irritation

No information is available on the effects of FB_1 on skin and eye irritation and/or sensitization.

7.1.4 Reproductive toxicity, embryotoxicity and teratogenicity

Concern about the reproductive and developmental effects of fumonisins originated with: (a) the observation of abortions in pregnant sows fed fumonisin-contaminated diets (Harrison et al., 1990); (b) the suggestion that a cluster of birth defects among residents in Brownsville, Texas, USA (Hendricks, 1999) might be associated with consumption of maize from the 1989 maize crop; (c) the association of "mystery swine disease" with fumonisin-contaminated maize (Bane et al., 1992); and (d) the discovery that fumonisins are inhibitors of sphingolipid biosynthesis (Wang et al., 1991). Currently there are no data to support the conclusion that consumption of fumonisins is a developmental or reproductive toxicant in farm animals or humans. There are also no data demonstrating that fumonisin consumption results in transfer to chicken eggs (Vudathala et al., 1994; Prelusky et al., 1996a) or that it crosses the placenta in rats (Voss et al., 1996a; Collins et al., 1998a,b), mice (Reddy et al., 1996) or rabbits (LaBorde et al., 1997).

Injection of purified FB_1 into fertile chicken eggs resulted in time- and dose-dependent embryopathic and embryocidal effects (Javed et al., 1993b). Embryonic changes included hydrocephalus, enlarged beaks and elongated necks. Pathological changes were noted in most organ systems. At the low FB_1 dose (1 μM = 0.72 $\mu g/ml$), stimulation of chick embryo development was observed. Stimulated embryo development *in vitro* in pre-somite rat embryos exposed to 0.5–1 $\mu g/ml$ of hydrolysed FB_1 has been reported in an abstract (Flynn et al., 1994). Higher concentrations of fully hydrolysed FB_1 (Flynn et al., 1997) and all concentrations of FB_1 ≥ 0.2 $\mu g/ml$ inhibited growth and development of pre-somite rat embryos *in vitro* (Flynn et al., 1994, 1996). Johnson et al. (1993) reported that FB_1 was a weak developmental toxin to organogenesis stage rat embryos (day 10.5; lowest-observed-effect level = 0.5 mM). FB_1 (≥ 2.5 mM = 1.8 $\mu g/ml$) inhibited reaggregation and growth of chicken embryo neural retina cells, a commonly used *in vitro* assay for screening potential developmental toxins (Bradlaw et al., 1994). Bacon et al. (1995) found effects of FB_1 in fertile chicken eggs similar to those reported by Javed et al. (1993b). In addition it was found that co-injection of fusaric acid and FB_1 resulted in a synergistic toxic response (Bacon et al., 1995). Zacharias et al. (1996) found that morphological changes, due to direct administration of FB_1 to chick embryos, were correlated with inhibition of glycosphingolipid biosynthesis.

Syrian hamsters orally gavaged with aqueous extracts of *F. verticillioides* (M 1325 = MRC 826) culture material containing fumonisins (0.25–18 mg FB_1/kg body weight) or pure FB_1 (12 mg/kg and 18 mg/kg) did not exhibit maternal toxicity based on weight gain, serum aspartate aminotransferase activity or total bilirubin. Histological examination of liver, kidney and placenta did not reveal important changes, although mild karyomegalic changes in liver were observed in the hamsters dosed with either aqueous extracts or pure FB_1 at > 6 mg FB_1/kg body weight (Floss et al., 1994a,b). When aqueous extracts were given by oral gavage from day 8 to day 10 or 12 of gestation, there appeared to be an increase in the number of fetal deaths, but statistical significance was not achieved (Floss et al., 1994a). Relative to controls, statistically significant increases in fetal deaths occurred only in the hamsters given 18 mg FB_1/kg body weight (aqueous culture extracts and pure material) (Floss et al., 1994b). Prenatal exposure to aqueous culture extracts containing fumonisins or to pure FB_1 were detrimental to fetal hamster survivability in the absence of maternal toxicity (Floss et al., 1994a,b; Penner et al. 1998).

In Fischer-344/N rats dosed orally from day 8 to 12 of gestation with 30 or 60 mg purified FB_1/kg body weight, the high dose significantly suppressed growth and fetal bone development while an extract of *F. proliferatum* (M 5991) in corn culture did not (Lebepe-Mazur et al., 1995a). Voss et al. (1996a) formulated diets using *F. verticillioides* (MRC 826) culture material to provide 0, 1, 10 or 55 mg FB_1/kg diet. Based on consumption, the diet containing 55 mg/kg provided about 3 to 4 mg FB_1/kg body weight per day to the dams. The diets were fed to male and female Sprague-Dawley rats prior to and during the mating, gestational and lactational phases of the study. Nephropathy was observed in males and females fed diets containing $\geq$ 10 mg/kg and 55 mg/kg, respectively. No statistically significant reproductive effects were observed in any of the males or females, and no developmental effects were found in fetuses during any phase of the study. Litter weight gains in the 10 and 55 mg/kg diet groups were slightly decreased. Increased levels of free sphinganine, a biomarker for fumonisin exposure, were demonstrated in the livers of dams in the 55 mg/kg diet group on gestation day 15. In contrast, no increase in the sphinganine/sphingosine (Sa/So) ratio was observed in fetuses at that time, suggesting that fetuses were not exposed *in utero* to FB_1. This finding was supported by the study in which an intravenous injection of $[^{14}C]FB_1$ was given to dams on gestation day 15. Radiolabel was easily detected in tissues of pregnant females but was not detected in their fetuses. Culture material containing fumonisins, and by inference FB_1, did not have reproductive effects at doses that were minimally toxic (Voss et al., 1996a). These findings have been recently confirmed (Sa/So ratios in fetuses were not affected and FB_1 was not teratogenic at the doses tested) in a Charles River CD rats (Collins et al., 1998a,b).

Gross et al. (1994) gavaged pregnant CD1 mice daily between gestation days 7 and 15 with a diet containing partially purified FB_1 extracted from *F. verticillioides* (M 1325 = MRC 826) culture material. Maternal toxicity and fetal developmental abnormalities (e.g., hydrocephalus, digital and sternal ossification) occurred at FB_1 dosages greater than 12.5 mg/kg body weight per day. Similar results were obtained in a second study using purified FB_1 (Reddy et al., 1996). As in the study by Voss et al. (1996b), the Sa/So ratio was significantly increased in maternal liver but not in fetal liver, suggesting that developmental effects were mediated through maternal toxicity (Reddy et al., 1996).

Unlike CD1 mice and Syrian hamsters, pregnant New Zealand white rabbits are very sensitive to the toxic effects of FB_1 (LaBorde et al., 1997). Maternal toxicity was observed at daily gavage dosages (in water) as low as 0.25 mg/kg body weight from gestational day 3 to gestational day 19. Compared to controls there was no increase in fetal loss or in gross visceral or skeletal abnormalities, and no decrease in fetal weight or fetal organ weight at any dosage (0 to 1.75 mg/kg body weight) (LaBorde et al., 1997). The maternal kidney, serum and urine Sa/So ratios were increased, but there were no increases in these ratios in fetal liver, brain or kidney (LaBorde et al., 1997). While FB_1 is toxic in the pregnant dam, it is not a developmental toxin but is maternally toxic in rabbits (LaBorde et al., 1997). However, the lowest-observed-effect level for maternal toxicity was 0.1 mg FB_1/kg body weight, which is equivalent to a calculated dietary fumonisin level of 2.3 mg/kg diet (LaBorde et al., 1997). Thus, in sensitive species, maternal toxicity and consequent fetal toxicity could occur at low dosages of FB_1.

There is currently no evidence of neonatal toxicity. However, average mean litter weights were reduced in litters from Sprague-Dawley dams fed *F. verticillioides* (MRC 826) culture material containing 10 or 55 mg FB_1/kg (Voss et al., 1996a). The Sa/So ratio was increased in litters at lactation day 21. However, given the likelihood that offspring had consumed the contaminated diets (Voss et al., 1996a), the authors could not ascertain the route of exposure (via milk or diet). Reduced weights and several alterations in haematological parameters were reported in mink kits lactationally exposed to fumonisins (Powell et al., 1996).

No FB_1 was detected in the milk of lactating sows fed diets containing non-lethal levels of FB_1 and there was no evidence of toxicosis in their suckling pigs (Becker et al., 1995). However, in a study with lactating cows administered FB_1 intravenously, the carry-over rate of FB_1 into the milk reached a maximum of 0.11% (Hammer et al., 1996), while in other studies no fumonisins were detected in cow's milk (Scott et al., 1994; Richard et al., 1996). In a reproductive study with mink, fumonisins were detected in the milk at 0.7% of the dietary fumonisin concentrations (Powell et al., 1996).

The question of neonatal toxicity is of concern since neonates may be more sensitive to fumonisins than adults. For example, a recent report by Kwon et al. (1997b) indicated that subcutaneous injection of FB$_1$ in neonatal rats caused elevation in the Sa/So ratio in brain tissue and reduced myelin deposition. The elevated sphinganine level was determined to be the result of a direct effect on the neonate brain, indicating that FB$_1$ can cross the blood-brain barrier (Kwon et al., 1997a). When maternal toxicity was minimal, there was little or no evidence of neonatal toxicity in rats (Ferguson et al., 1997).

7.1.5 *Mutagenicity and related end-points*

The fumonisins FB$_1$, FB$_2$ and FB$_3$ (98, 98, 90% pure, respectively) were non-mutagenic in the *Salmonella* assay against the tester strains TA97a, TA98, TA100 and TA102, in both the presence and absence of the S-9 microsomal preparation (Gelderblom & Snyman, 1991). The non-mutagenicity of FB$_1$ (approximately 90% pure) to *Salmonella* tester strain TA100 at concentrations up to 100 mg/plate was confirmed by Park et al. (1992). Similarly negative results were reported with FB$_1$ in *Salmonella* TA98 and TA100, as well as in SOS chromotest in *E. coli* PQ37 and differential DNA repair assays with *E. coli* K12 strains (343/753, *uvr*B/*rec*A and 343/765, *uvr⁺rec⁺*) (Knasmüller et al., 1997). In contrast, Sun & Stahr (1993), using a commercial bioluminescent bacterial (*Vibrio fischeri*) genotoxicity test, reported that FB$_1$ showed in the concentration range 5–20 µg/ml genotoxic activity without the metabolic activation of S-9 fraction.

FB$_1$ (and FB$_2$) were non-genotoxic in the *in vitro* rat hepatocyte DNA repair assay at concentrations ranging from 0.04 to 80 µM (and FB$_2$ from 0.04 to 40 µM) as well as in the *in vivo* assay at a dose of 100 mg/kg body weight administered by gavage (FB$_1$ or FB$_2$) (Gelderblom et al., 1989, 1992b). The finding that FB$_1$ does not induce unscheduled DNA synthesis was confirmed in the *in vitro* assay in primary rat hepatocytes at concentrations ranging from 0.5 to 250 µM (Norred et al., 1992a).

FB$_1$ induced DNA strand breaks in isolated rat liver nuclei (Sahu et al., 1998).

FB$_1$ induced a moderate increase in the micronucleus frequency in primary rat hepatocytes at concentrations ranging from 0.01 to 1 µg/ml. No concentration-dependent increase of micronuclei occurred. A

41

significant concentration-dependent increase in chromosomal aberrations was observed in isolated hepatocytes exposed to FB_1 at concentrations ranging from 1 to 100 µg/ml (Knasmüller et al., 1997).

FB_1 induced lipid peroxidation in isolated rat liver nuclei at concentrations ranging from 40 to 300 µM (Sahu et al., 1998). It also increased significantly the level of thiobarbituric-acid-reactive substances *in vitro* in primary rat hepatocytes at concentrations of 75 and 150 µM and *in vivo* in the liver of rats fed a dietary FB_1 level of 250 or 500 mg/kg for 21 days (Abel & Gelderblom, 1998).

Single gavage doses of 50, 100 and 200 mg FB_1/kg body weight significantly ($P = 0.05$) inhibited hepatocyte proliferation as measured by the incorporation of radiolabelled thymidine into DNA in partially hepatectomized male Fischer rats (Gelderblom et al., 1994). Inhibition of hepatocyte proliferation was also observed after dietary exposure to FB_1 ($\geq$ 50 mg/kg diet) (Gelderblom et al., 1996c). FB_1 also inhibited DNA synthesis induced by epidermal growth factor in primary rat hepatocytes (Gelderblom et al., 1995).

In BALB/3T3 A31-1-1 mouse embryo cells, FB_1 (90% pure) treatment produced transforming activity at 500 µg/ml but not at lower or higher concentrations (Sheu et al., 1996).

7.1.6 Carcinogenicity

7.1.6.1 Carcinogenicity bioassays

When inbred BD IX rats were fed commercial diet containing freeze-dried or oven-dried culture material inoculated with *F. verticillioides* MRC 826 for 2 years, the incidence of liver tumours (hepatocellular and cholangiocellular carcinomas combined) was increased (control: 0/20, freeze-dried 13/20, oven-dried, 16/20) (Marasas et al, 1984b). When 30 rats of the same inbred strain were given *F. verticillioides* MR 826 (containing fusarin C and later found to produce FB_1 and FB_2) for 23-27 months, two hepatocellular and eight cholangiocellular cancers were observed. In addition, neoplastic hepatic nodules were observed in all surviving 21 animals, but none among the controls. However, in rats similarly administered *F. verticillioides* MRC 1069 culture material containing 104 mg/kg fusarin C (but suspected of being low in FB_1), no increase in hepatic carcinomas was observed (Jaskiewicz et al., 1987b).

Maize from a field outbreak of equine leukoencephalomalacia in the USA, shown to be naturally contaminated with *F. verticillioides*, was fed to 12 male Fischer-344 rats and commercial rodent feed to 12 controls by Wilson et al. (1985). All treated rats necropsied from 123 to 176 days had multiple hepatic neoplastic nodules, adenofibrosis and cholangiocarcinoma, whereas no such lesions were found in the controls. The authors considered these lesions in the livers of male Fischer-344 rats to be similar to those described in male BD IX rats by Marasas et al. (1984b). The fact that the lesions observed by Wilson et al. (1985) developed more rapidly (as early as 123 days) than those described by Marasas et al. (1984b) (more than 450 days) was attributed by Wilson et al. (1985) to the dietary deficiencies, which included choline and methionine, in the maize-only diet used in their study.

Hendrich et al. (1993) reported that in Fischer-344/N rats fed diets containing *Fusarium proliferatum* maize culture material (with known amounts of FB_1) there was an increased incidence of hepatocellular adenomas, relative to rats fed the control diets. When rats were fed nixtamalized *Fusarium proliferatum* (M 5991) maize culture material diets (converting FB_1 to hydrolysed FB_1) the incidence of hepatic adenomas and cholangiomas was reduced relative to rats fed the diets containing FB_1. The frequency of hepatic/cholangiocellular adenomas in rats given the nixtamalized diet was higher in rats receiving nutrient-supplemented diet (equivalent to AIN-76) than in rats given a diet not supplemented with nutrients (nutritionally deficient relative to AIN-76). In a 4-week feeding study with male Sprague-Dawley rats, Voss et al. (1996c) found that hydrolysed FB_1 (58 mg/kg diet) from nixtamalized *F. verticillioides* (MRC 826) was both hepatotoxic and nephrotoxic. However, the extent and severity of the hepatotoxicity was significantly less than that caused by FB_1 (71 mg/kg diet) from *F. verticillioides* (MRC 826), whereas the kidney toxicity was similar (Voss et al., 1996c).

In male BD IV rats treated with the known oesophageal carcinogen *N*-methylbenzylnitrosamine (NMBA) (2.5 mg/kg body weight) and FB_1 (5 mg/kg body weight), there was no synergistic interaction between

NMBA and FB_1 in the rat oesophagus when the two compounds were administered together (Wild et al., 1997).

A semi-purified maize-based diet containing FB_1 (not less than 90% pure) at 50 mg/kg diet was fed to 25 inbred male BD IX rats over a period of 26 months (Gelderblom et al., 1991). A control group received the same diet without FB_1 (the FB_1 content of the control diet was approximately 0.5 mg/kg and no aflatoxin B_1 could be detected). Five rats from each group were killed at 6, 12, 20 and 26 months. All FB_1-treated rats (50 mg/kg diet) that died or were killed from 18 months onward suffered from a micro- and macronodular cirrhosis and had large expansile nodules of cholangiofibrosis at the hilus of the liver. The pathological changes terminating in cirrhosis and cholangiofibrosis were already present in the liver of rats killed 6 months after the initiation of the experiment and included fibrosis, bile duct hyperplasia and lobular distortion. The severity of the hepatic lesions increased with time and the histological changes were consistent with those of a chronic toxic hepatosis progressing to cirrhosis. Ten out of 15 FB_1-treated (50 mg/kg diet) rats (66%) – but none in the controls – that were killed or died between 18 and 26 months developed primary hepatocellular carcinoma. Metastases to the heart, lungs or kidneys were present in four of the rats with hepatocellular carcinoma. Apart from the hepatocellular carcinoma, FB_1 also induced cholangiofibrosis consistently from 6 months onward, and toward the end of the experiment, cholangiocarcinoma. However, the authors noted that the experiment was performed under nutritionally compromised conditions, using diets deficient in vitamins, methionine and choline, that may have had an enhancing effect on the action of FB_1 in the liver (Table 3).

The detailed results of a second long-term experiment in rats fed diets containing 0, 1, 10 and 25 mg FB_1/kg diet over a period of 24 months have not yet been published (Gelderblom et al., 1996b). However, in preliminary reports, Gelderblom et al. (1996b, 1997) noted that no cancers were observed in these rats, including those fed 25 mg FB_1/kg diet.

Male and female Fischer-344/N Nctr BR rats and B6C3F₁/Nctr BR mice were given diets containing FB_1 for 2 years as part of the US National Toxicology Program (NTP) tumorigenesis studies on FB_1

Table 3. Summary of the induction of neoplasia in long-term feeding studies[a]

| Neoplasia | Species and strain | Sex | Fumonisin concentration (mg/kg feed) | | | | | | | Reference |
			0	5	15	50	80	100	150	
Hepatocellular carcinoma	BD IX rats	male	0/15			10/15			5/48	Gelderblom et al. (1991)
Cholangiofibrosis[b]	BD IX rats	male	0/15			15/15			10/48	
Renal tubule adenoma	F-344/N Nctr rats	male	0/48	0/40	0/48	2/48				US NTP (1999)
Renal tubule carcinoma	F-344/N Nctr rats	male	0/48	0/40	0/48	7/48				US NTP (1999)
Renal tubule adenoma	F-344/N Nctr rats	female	0/48	0/40	1/48	0/48		0/48		US NTP (1999)
Renal tubule carcinoma	F-344/N Nctr rats	female	0/48	0/40	0/48	0/48		1/48		US NTP (1999)
Hepatocellular adenoma	B6C3F$_1$/Nctr mice	female	5/47	3/48	1/48	16/47	31/45			US NTP (1999)
Hepatocellular carcinoma	B6C3F$_1$/Nctr mice	female	0/47	0/48	0/48	10/47	9/45			US NTP (1999)
Hepatocellular adenoma	B6C3F$_1$/Nctr mice	male	9/47	7/47	7/48		6/48		8/48	US NTP (1999)
Hepatocellular carcinoma	B6C3F$_1$/Nctr mice	male	4/47	3/47	4/48		3/48		2/48	US NTP (1999)

[a] This summarizes the data for tissues where fumonisin dose-dependent induction of tumours was detected (US NTP, 1999)

[b] In Gelderblom et al. (1991), cholangiofibrosis was considered to have progressed to cholangiocarcinoma

(US NTP, 1999). The FB_1 that was used in these studies was > 96% pure. Dietary levels of FB_1 were 0, 5, 15, 50 and 150 mg/kg diet for the male rats, resulting in average daily FB_1 doses of 0, 0.3, 0.9, 3.0 and 6.0 mg/kg body weight. The dietary levels for female rats were 0, 5, 15, 50 and 100 mg/kg diet, resulting in average daily FB_1 doses of 0.25, 0.8, 2.5 and 7.5 mg/kg body weight.

There was no difference in body weight, survival or feed consumption in male rats fed FB_1 when compared to rats on control diets. The only compound-related change in tumour incidence was the induction of renal adenomas and carcinomas in male rats. The overall incidence of renal tubule tumours in male rats receiving 0, 5, 15, 50 and 150 mg FB_1/kg diet were 0/48, 0/40, 0/48, 2/48 and 5/48 for adenomas and 0/48, 0/40, 0/48, 7/48 and 10/48 for carcinomas. Renal tubule adenomas were characterized as an expansive proliferation of renal tubule epithelial cells that tended to be separated into lobules by a delicate fibrous stroma. The neoplastic cells had nuclei that were slightly larger with increased cytoplasmic volume. The cytoplasmic changes were uniform within individual lesions and varied from clear to basophilic. Renal tubule carcinomas were characterized by cellular atypia, necrosis within a lesion, invasion of the adjacent normal renal parenchyma, or metastasis to distant organs. Historically, renal tubule carcinomas have not occurred in Fischer-344/N Nctr male rats on control diets. There was no apparent involvement of α-2 microglobulin in the tumorigenicity in male rat kidneys. The incidence in renal tumours was accompanied by an increased incidence in renal tubule epithelial cell hyperplasia at 50 and 150 mg FB_1/kg diet at 2 years (2/48, 1/40, 4/48, 14/48 and 8/48 of the male rats receiving 0, 5, 15, 50 and 150 mg FB_1/kg diet). Similarly, increased renal tubule epithelial cell apoptosis and proliferation were detected at 50 and 150 mg FB_1/kg diet in male rats sacrificed following 6, 10, 14 and 26 weeks on FB_1-containing diets. As in the 28-day range-finding studies, the apoptosis was confined to tubules of the inner cortex and was characterized by cellular shrinkage from adjacent cells, cytoplasmic eosinophilia, and chromatin condensation and margination in the nucleus. Apoptotic cells were additionally detected using an *in situ* method for detection of DNA fragmentation. In serum removed from male rats killed at 6, 10, 14 or 26 weeks, no FB_1-dependent changes were noted (e.g., in cholesterol, triglycerides or serum alanine aminotransferase levels). The urinary Sa/So ratio was increased in the urine of male rats fed

5, 15, 50 or 150 mg FB_1/kg diet, while kidney tissue Sa/So ratios were increased at 15, 50 or 150 mg FB_1/kg diet. Increased tissue sphingoid base changes, and renal tubule tumour incidence and hyperplasia were detected at 50 and 150 mg FB_1/kg diet, while urinary sphingolipid changes were detected at 15, 50 or 150 mg FB_1/kg diet. In the livers of the male rats, an increase in basophilic foci was detected at 150 mg FB_1/kg diet, but liver tissue Sa/So ratios were not affected.

FB_1 at 5, 15, 50 and 100 mg/kg diet did not affect body weight, survival, feed consumption, serum analytes or tumour incidence in the female rats. Significant increases in urinary and kidney tissue Sa/So ratios were detected at 50 and 100 mg FB_1/kg diet, but the extent of induction was not as high as in the male rats.

In the female mice, dietary levels of FB_1 were 0, 5, 15, 50 and 80 mg/kg diet, resulting in average daily FB_1 consumption of 0.7, 2.1, 7.0 and 12.5 mg/kg body weight, respectively. In the male mice, dose levels of 0, 5, 15, 80 and 150 mg FB_1/kg diet resulted in average daily FB_1 doses of 0, 0.6, 1.7, 9.5 and 17 mg/kg body weight, respectively.

In female $B6C3F_1$/Nctr mice, there were essentially no differences in the mean body weights and diet consumption. The body weights of the mice on this study were less than those reported for the NTP and control studies at the National Center for Toxicological Research (NCTR). This was attributed to an unintended restriction of the powdered feed in the individual feeders. The female mice consuming diets containing 80 mg FB_1/kg diet had a significantly reduced survival compared to the female mice on control diets.

The only tissue that demonstrated FB_1-dependent changes in tumour incidence was the liver in the female mice. Hepatocellular adenomas were present in 5/47, 3/48, 1/48, 16/47 and 31/45 female mice and hepatocellular carcinomas were present in 0/47, 0/48, 0/48, 10/47 and 9/45 female mice consuming 0, 5, 15, 50 and 80 mg FB_1/kg diet, respectively. Hepatocellular adenomas were characterized as discrete lesions with compression of adjacent normal tissues. The normal hepatic lobular structure was absent with uneven growth patterns. The cells in the adenoma appeared to be well differentiated and either eosinophilic, basophilic or vacuolated. Hepatocellular carcinomas were characterized as foci of cells with distinct trabecular or adenoid structure. Histological evidence of local invasiveness or metastasis was usually evident. The cells within the carcinoma were poorly differentiated or anaplastic. Some of the carcinomas appeared to arise within adenomas. The incidence of hepatocellular adenomas and carcinomas in the female mice was within the range of historical occurrence in B6C3F₁/Nctr mice. The increased tumour incidence was accompanied by increased hepatocellular hypertrophy (0/47, 0/48, 0/48, 27/47, 31/45) and hepatocellular apoptosis (0/47, 0/48, 0/48, 7/47, 14/45) in the female mice. The mean liver weights (relative to body weight) of the mice after 2 years were increased at 50 and 80 mg FB₁/kg diet. There were no consistent increases in serum analytes in mice killed at 3, 7, 9 and 24 weeks. Liver sphingoid bases were increased at 80 mg FB₁/kg diet at weeks 3, 7 and 9 but not week 24. Levels of urinary sphingoid bases were not determined.

No FB₁-dependent changes in tumour incidences in the male mice receiving diets of 0, 5, 15, 80 and 150 mg FB₁/kg diet were identified. The body weights, organ weights, survival, feed consumption and serum analytes were not affected by FB₁ dose. The lack of demonstration of statistically significant sphingoid base increases in livers at intermediate sacrifices (3, 7 and 9 weeks) were probably due to the low sample number (n = 4) (US NTP, 1999).

7.1.6.2 Short-term assays for carcinogenicity

The short-term assays have mainly used rat liver nodules as the end-point, assessed either using traditional microscopy or histochemical analysis of different enzyme activities such as gamma-glutamyl transferase (GGT) or placental glutathione *S*-transferase (PGST). In many of these assays, different stages (initiation, promotion) of the multistage carcinogenesis model have been investigated by combining the FB₁ treatment with classical promotion assays. In these studies FB₁ (or a *Fusarium* extract) was administered alone, or before a promoting treatment, or after an initiating treatment.

a) *Initiation studies*

In male BD IX rats fed a diet containing 0.1% FB₁ during 4 weeks, GGT-positive (GGT⁺) foci were induced in the liver (Gelderblom et al., 1988).

GGT$^+$ foci were induced in the liver of male Fischer rats fed a diet containing 0.5–1 g/kg of FB$_1$ (90–95% pure) for 21–26 days, followed by partial hepatectomy and treatment with 2-AAF and carbon tetrachloride. However, foci were not induced when single or multiple doses (50–200 mg/kg) of FB$_1$ (and FB$_2$) were administered by gavage to hepatectomized rats (Gelderblom et al., 1992b, 1993).

In subsequent dose–response studies in male Fischer rats using the same experimental approach, Gelderblom et al. (1994) reported that the lowest dietary level to produce cancer initiation (GGT$^+$-foci) over 21 days was 250 mg FB$_1$/kg diet. The lowest levels to cause cancer initiation over 14 and 7 days were 500 mg and 750 mg FB$_1$/kg diet, respectively. Based on the feed intake values, the effective dosage level (EDL) for cancer initiation over a period of 21 days was 142 < EDL < 308 mg FB$_1$/kg body weight and over 14 days the amount required for cancer initiation was 233 < EDL < 335 mg FB$_1$/kg body weight. The dietary level of FB$_1$ required for cancer initiation is dependent on the duration of exposure since a dose of 293 mg FB$_1$/kg body weight over 7 days did not initiate cancer whereas a similar dose (308 mg FB$_1$/kg body weight) over 21 days did.

Lebepe-Mazur et al. (1995b) fed female Fischer-344/N rats for one week with a semipurified diet with or without an aqueous extract of a *Fusarium verticillioides* (M 1325 = MRC 826) culture, providing 20 mg FB$_1$/kg diet, and administered a single dose of 30 mg/kg body weight diethylnitrosamine (DEN) thereafter. Rats fed the *Fusarium* culture showed more PGST$^+$ hepatocytes than those treated with DEN alone. Continued dietary treatment with the *Fusarium moniliforme* culture for 12 weeks after the DEN administration did not further increase the number of PGST foci. When the FB$_1$ diet (one week) was followed by the DEN administration and a 7-day-non-treatment interval, no increase in PGST foci by FB$_1$ was observed.

In another study (Lebepe-Mazur et al., 1995c), female Sprague-Dawley rats were fed a diet supplemented with corn contaminated with *Fusarium proliferatum* (containing 20 or 50 mg/kg FB$_1$) for six months; GGT-foci were not observed but the number of PGST$^+$ altered hepatic foci was increased in treated rats in comparison to rats fed a semipurified diet without supplementation. In a similar study (Lebepe-Mazur et al., 1995b), feeding diet containing 20 mg/kg FB$_1$ for one or

13 weeks failed to induce a statistically significant increase in PGST-altered foci.

In male Sprague-Dawley rats administered purified FB_1 intraperitoneally at 10 mg/kg body weight per day for 4 days, as well as in male and female Sprague-Dawley rats given 35 and 75 mg/kg body weight per day orally for 11 days, significant increases in $PGST^+$ hepatocytes were observed (Mehta et al., 1998).

b) *Promotion studies*

In a study on the promotion activity of FB_1, it was administered to male Fischer-344 rats (10, 50, 100, 250 or 500 mg/kg diet for 21 days) after a dose of DEN (200 mg/kg body weight) (Gelderblom et al., 1996c). Dietary levels of 50 mg/kg or more markedly increased the number and size of the $PGST^+$ foci in the liver. It was thus concluded (Gelderblom et al., 1996b,c) that the dose of FB_1 required for cancer initiation was markedly higher than that required for cancer promotion.

Female Sprague-Dawley rats were fed a semipurified diet with or without an aqueous extract of a *Fusarium verticillioides* (M 1325 = MRC 826) culture, providing 20 or 50 mg FB_1/kg for 6 months, after a single dose of 30 mg DEN/kg. The number of PGST-altered foci was increased at 20 mg/kg but not in the high-dose group, as compared to rats treated with DEN alone. GGT-altered foci were not observed (Lebepe-Mazur et al., 1995c).

Gelderblom et al. (1996b) suggested that FB_1-induced hepato-carcinogenesis in male BD IX rats developed against a background of chronic toxic hepatosis culminating in cirrhosis. Chronic hepatotoxicity appears to be a prerequisite for the development of liver cancer in the BD IX rat (Gelderblom et al., 1996b).

7.2 Other mammals

7.2.1 Equine leukoencephalomalacia

Equine leukoencephalomalacia (ELEM) syndrome is characterized by the presence of liquefactive necrotic lesions in the white matter of the cerebrum. The name is somewhat misleading since the gray matter may also be involved (Marasas et al., 1988a). This fatal disease apparently occurs only in equids, although there has been one unconfirmed report of fumonisin-induced brain lesions and haemorrhage in rabbits gavaged with FB_1 (Bucci et al., 1996) and there is some evidence that FB_1 can cross the blood-brain barrier and disrupt brain sphingolipid metabolism in neonatal rats (Kwon et al., 1997b). In equids, the ELEM syndrome has been recognized since the 19th century as a sporadically occurring condition. ELEM was experimentally produced by feeding mouldy maize obtained from a field case in Kansas by Butler (1902). The disease was known as "mouldy maize poisoning" but attempts to identify the responsible fungus failed.

Wilson & Maronpot (1971) succeeded in establishing the causative agent when they isolated *F. verticillioides* as the predominant contaminant of mouldy maize that had caused cases of ELEM in Egypt and reproduced ELEM by feeding culture material of the fungus on maize to two donkeys. Subsequently investigators in South Africa confirmed the ability of *F. verticillioides* (MRC 826) culture material to induce the characteristic clinical signs and pathological changes of ELEM as well as hepatosis in horses and donkeys (Kellerman et al., 1972; Marasas et al., 1976, 1988a; Kriek et al., 1981).

The first symptoms of the syndrome are lethargy, head pressing and inappetence, followed by convulsions and death after several days. Elevated serum enzyme levels indicative of liver damage (Wilson et al., 1992) are preceded by elevation in the serum Sa/So ratio (Wang et al., 1992; Riley et al., 1997). Serum enzyme levels often return to near normal concentrations (Wang et al., 1992; Wilson et al., 1992; Ross et al., 1993; Riley et al., 1997) but usually increase markedly immediately prior to or at the onset of behavioral changes (Kellerman et al., 1990; Wang et al., 1992; Ross et al., 1993; Riley et al., 1997).

In addition to the brain lesions, histopathological abnormalities in liver and kidney have been reported in horses orally dosed with pure fumonisins, maize screenings naturally contaminated with fumonisins, or culture material containing known amounts of fumonisins (Kellerman et al., 1990; Wilson et al., 1992; Ross et al., 1993; Caramelli et al., 1993).

Shortly after the isolation and structure elucidation of fumonisins in 1988 (Bezuidenhout et al., 1988; Gelderblom et al., 1988), Marasas et al. (1988a) successfully produced ELEM in a horse by the intravenous administration of pure FB_1. This was done by avoiding as much as possible hepatotoxicity using serum enzymes indicative of it. ELEM has also been produced in horses given pure FB_1 by stomach tube, again monitoring for liver toxicity (Kellerman et al., 1990).

Fatal liver disease in the absence of any brain lesions has been induced by intravenous injection of FB_1 (Laurent et al., 1989b). ELEM concurrent with significant liver disease has been observed in horses and ponies fed feeds naturally contaminated with fumonisins at low concentrations (Wilson et al., 1992; Ross et al., 1993). The development of brain lesions in the absence of major liver lesions does not preclude biochemical dysfunction in non-brain tissue from contributing to the brain lesions. Ross et al. (1993) concluded that length of exposure, level of contamination, individual animal differences, previous exposure, or pre-existing liver impairment may all contribute to the appearance of the clinical disease.

To date, the lowest FB_1 dose that has resulted in ELEM, in a controlled experiment, is 22 mg/kg in diets formulated with naturally contaminated maize screenings (Wilson et al., 1992). Analysis of feeds from confirmed cases of ELEM indicated that consumption of feed with a FB_1 concentration greater than 10 mg/kg diet is associated with increased risk of development of ELEM, whereas, a concentration less than 6 mg/kg diet is not (Ross, 1994).

A study by the National Veterinary Services Laboratory of the US Animal and Plant Health Inspection Agency (National Veterinary Services Laboratory, 1995) showed that horses fed 15 mg FB_1/kg in diets formulated from *F. proliferatum* (M 5991) culture material did not exhibit any clinical signs or altered serum biochemical parameters (including changes in the Sa/So ratio) after 150 days. A similar result was found with a pony fed a diet containing maize screenings naturally contaminated with 15 mg FB_1/kg (Wang et al., 1992). Thus, the minimum toxic dose in equids appears to be < 22 mg/kg > 15 mg/kg based on studies with naturally contaminated maize screenings or culture material (*F. proliferatum*) containing fumonisins. The minimum toxic dose of pure fumonisins is unknown.

In a study using culture material containing primarily FB_2 or FB_3, Ross et al. (1994) found that a diet formulated from *F. proliferatum* (M 6290 and M 6104) culture material containing primarily FB_2 at 75 mg/kg was capable of inducing ELEM with hepatic involvement in ponies after 150 days. In contrast, diets containing primarily FB_3 (75 mg/kg) were without any effect (serum enzymes, clinical signs and histology were all normal relative to control ponies) after 57 to 65 days. It was concluded that FB_3 was less toxic than FB_2 or FB_1 (Ross et al., 1994). However, analysis of serum and tissues from ponies fed the FB_3 diets revealed that the FB_3 diets significantly increased concentrations of free sphingoid bases relative to controls and that serum enzymes were elevated but within the normal range for ponies (Riley et al., 1997).

7.2.2 *Porcine pulmonary oedema syndrome*

The first report of the disease now known as porcine pulmonary oedema (PPE) was by Kriek et al. (1981). In experimental trials, culture material of *F. verticillioides* (MRC 826) was fed to horses, pigs, sheep, rats and baboons (Kriek et al.,1981). Lung oedema occurred only in pigs. Clinical signs of PPE typically occur soon (2–7 days) after pigs consume diets (culture material or contaminated maize screenings) containing large amounts of fumonisins over a short period of time. Clinical signs usually include dyspnoea, weakness, cyanosis and death (Osweiler et al., 1992). At necropsy, the animals exhibit varying degrees of interstitial and interlobular oedema, with pulmonary oedema and hydrothorax (Colvin & Harrison, 1992; Colvin et al., 1993). Varying amounts of clear yellow fluid accumulate in the pleural cavity.

Toxic hepatosis occurs concurrently with PPE (Osweiler et al., 1992; Colvin et al., 1993) and is also observed in animals that consume high levels of fumonisins but do not develop PPE (Haschek et al., 1996). Typically, the liver contains multiple foci of coagulative necrosis that do not show zonal distribution across the three zones of the liver (Osweiler et al., 1992; Colvin et al., 1993). Two studies have reported nodular hyperplasia in the pig liver (Casteel et al., 1993, 1994).

The physiological alteration that results in the inability of the lung to maintain fluid equilibrium is unknown. However, several hypotheses have been proposed that are supported by experimentation. Casteel et al. (1994) found that feeding culture material diets (M 1325 = MRC 826) containing 150 to 170 mg FB_1/kg for 210 days resulted in right ventricular hypertrophy and medial hypertrophy of the pulmonary arterioles. It was suggested that this cardiotoxic effect was an indirect consequence of fumonisin-induced hepatotoxicity. Cardiac failure is a well-known physiological mechanism inducing altered pulmonary haemodynamics which can result in pulmonary oedema (Colvin et al., 1993). Significant changes in oxygen consumption and several haemodynamic parameters in pigs fed diets containing fumonisins suggest that pulmonary hypertension caused by hypoxic vaso-constriction may contribute to PPE (Smith et al., 1996a,b). It has been hypothesized that the cardiovascular alterations are a consequence of sphingoid-base-induced inhibition of L-type calcium channels (Smith et al., 1996b).

Haschek et al. (1992) hypothesized that PPE might be induced by dysfunction of pulmonary interstitial macrophages (PIM) resulting in release of vasoactive mediators. The accumulation of membranous materials in PIM, secondary to hepatotoxicity, was postulated as the possible basis for PIM dysfunction (Haschek et al., 1992). A similar phenomena has been observed in alveolar endothelial cells (Gumprecht et al., 1998). It has been shown that consumption of culture material diets (MRC 826) containing fumonisins does in fact alter PIM function (Smith et al., 1996c). How this might contribute to pulmonary oedema is not clear. However, it has been hypothesized that PIM dysfunction could contribute to increased susceptibility to microbial diseases (Smith et al., 1996c). It has been shown that serum tumour necrosis factor-α (TNF-α)-like activity was increased in pigs fed culture material (M 1325 = MRC 826) containing 150 mg FB_1/kg (Guzman et al., 1997). Fumonisin-induced changes in the TNF pathway have also been seen in lipopolysaccharide-stimulated macrophages collected from BALB/c mice dosed with pure FB_1 (Dugyala et al., 1998).

In 1989-1990 outbreaks of this disease were reported in different parts of the USA (Harrison et al., 1990; Osweiler et al., 1992; Ross et al., 1992). Maize screenings obtained from farms (Harrison et al., 1990; Osweiler et al., 1992) where pigs died of PPE were predominantly contaminated with *F. verticillioides.* Feeding (Kriek et al., 1981; Osweiler et al., 1992; Fazekas et al., 1998) or intubation (Colvin et al., 1993) of *F. verticillioides* culture material (MRC 826) produces PPE. Also, PPE and hepatotoxicity have been produced by feeding diets containing maize screenings naturally contaminated with fumonisins (Osweiler et al., 1992; Motelin et al., 1994). Purified FB_1 has been shown to produce the disease when administered intravenously (Harrison et al., 1990; Haschek et al., 1992; Osweiler et al., 1992). However, PPE has not yet been produced by oral administration of pure fumonisins.

As with ELEM, there is a strong correlation between fumonisin content of maize screenings obtained from different farms and outbreaks of PPE (Osweiler et al., 1992; Ross et al., 1992; Ross, 1994). The highest concentration of FB_1 ever reported was from maize screenings (330 mg/kg) associated with an outbreak of PPE (Ross et al., 1992). The minimum toxic dose has not been clearly established. Osweiler et al. (1992) induced PPE by feeding *F. verticillioides* (MRC-3033) maize culture material reportedly containing 17 mg FB_1/kg for 5 days. In the same study, maize screenings containing fumonisins at 92 mg/kg induced PPE in several pigs after 5–7 days; similar results were obtained in studies by Harrison et al. (1990), Haschek et al. (1992) and Motelin et al. (1994). Based on feeding studies with maize screenings naturally contaminated with fumonisins, FB_1 concentrations of 92 to 166 mg/kg have induced PPE in 4–7 days.

Pigs fed diets containing fumonisins (formulated with culture material or naturally contaminated maize screenings) often do not die of PPE, even when fumonisins are reported to be present at very high concentrations in the diet. Concentrations of FB_1 as low as 17 mg/kg in culture material diets (MRC-3033) induced PPE in 5 days (Osweiler et al.,1992). In contrast, culture-material-formulated (M 1325 = MRC 826) diets containing as much as 190 mg/kg have been fed for 83 days with no reported evidence of respiratory distress (Casteel et al., 1993) and a dose of 150 to 170 mg/kg diet for up to 210 days caused liver effects early on but no evidence of pulmonary oedema (Casteel et al., 1994).

Colvin et al. (1993) concluded that the primary determinant of whether pulmonary oedema or liver failure caused death was the quantity of fumonisins fed or intubated per kg body weight per day.

They proposed that > 16 mg/kg body weight per day induced PPE and < 16 mg/kg body weight per day induced liver failure. However, daily oral intake levels of FB_1 plus FB_2 (maize screenings) from 4.5 to 6.3 mg/kg body weight have induced PPE (Haschek et al., 1992; Motelin et al., 1994). The FB_1 concentration in these diets was 166 mg/kg and 129 mg/kg, respectively. Liver lesions have been induced with maize screenings at 1.1 mg/kg body weight per day (17 mg FB_1/kg diet) (Motelin et al., 1994).

In pigs, tissues other than liver and lung have been reported to be targets for fumonisins, e.g., pancreas (Harrison et al., 1990), heart (Casteel et al., 1994), kidney (Colvin et al., 1993; Harvey et al., 1995, 1996), pulmonary intravascular macrophages (Haschek et al., 1992), and oesophagus (Casteel et al., 1993). None of these studies were conducted with pure fumonisins. In a recent study with pure FB_1, altered growth and changes in selected haematological parameters in pigs were reported at dietary levels as low as 1 mg/kg (Rotter et al., 1996).

7.2.3 Poultry toxicity

Several reports have been published implicating *F. verticillioides* contamination of feed in diseases of poultry (Marasas et al., 1984a; Bryden et al., 1987; Jeschke et al., 1987; Prathapkumar et al., 1997). The clinical features of the disease often include diarrhoea, weight loss, increased liver weight and poor performance. Immunosuppression in chickens was also produced in birds fed maize cultured with several different isolates of the fungus (Marijanovic et al., 1991). Functional and morphological changes were observed in chicken exposed to FB_1 (Qureshi & Hagler, 1992). Several studies have confirmed that *F. verticillioides*, *F. proliferatum*, FB_1 and moniliformin are toxic to poultry (broiler chicks, turkeys, ducklings) (Ledoux et al., 1992, 1996; Brown et al., 1992; Dombrink-Kurtzman et al., 1993; Javed et al., 1993a, 1995; Weibking et al., 1993a,b, 1995; Kubena et al., 1995a,b; Hall et al., 1995; Bermudez et al., 1996; Vesonder & Wu, 1998) and chicken embryos (Javed et al., 1993b; Bacon et al., 1995). The levels of fumonisins used in these studies were 75–644 mg/kg diet. Culture materials and naturally contaminated maize containing *F. proliferatum* may contain, in addition to fumonisins, moniliformin and beauvericin (Kriek et al., 1977; Logrieco et al., 1993; Plattner & Nelson, 1994).

Espada et al. (1994) reported toxicity and altered haematological parameters (Espada et al., 1997) in broiler chicks fed diets containing pure FB_1 (10 mg/kg) and FB_1(30 mg/kg) from *Fusarium verticillioides* (MRC 826) culture material.

7.2.4 Non-human primate toxicity

Kriek et al. (1981) fed three baboons *F. verticillioides* culture material (MRC 826). Baboon 1 and baboon 2 died of acute congestive heart failure after 248 and 143 days, respectively. The remaining baboon continued on feed for 720 days, at which time it was killed. Autopsy of baboon 3 revealed that the principle lesion was cirrhosis of the liver. Vervet monkeys fed *F. verticillioides* culture material (MRC 826) for 180 days exhibited various degrees of toxic hepatosis (Jaskiewicz et al., 1987a). Subsequent long-term studies (Fincham et al., 1992) with vervet monkeys fed MRC 826 culture material shown to contain fumonisins revealed an increase in serum cholesterol, plasma fibrinogen and blood coagulation factor VII (factors known to promote atherosclerosis). These changes occurred secondary to chronic hepatotoxicity at a dose calculated to average 0.3 mg total fumonisins/kg body weight per day (low-dose diet) based on a retrospective analysis of the diets (Fincham et al., 1992) and a high-dose diet averaging approximately 0.8 mg total fumonisins/kg body weight per day (Shephard et al., 1996b). Analysis of the free sphingoid bases in serum from some of the animals used in the study by Fincham et al. (1992) showed that in serum the free sphinganine concentration and Sa/So ratio were significantly elevated in both the low-dose and high-dose animals (Shephard et al., 1996b). Free sphinganine and the Sa/So ratio were also elevated in urine at both dose levels, but not significantly (Shephard et al., 1996b).

7.2.5 Other species

Other species that have been studied using pure fumonisins, contaminated maize screenings or maize culture material of *F. verticillioides* include the following: catfish (Brown et al., 1994; Goel et al., 1994); cattle (Osweiler et al., 1993); hamsters (Floss et al., 1994a,b); lambs (Edrington et al., 1995); mink (Restum et al., 1995); and rabbits (Gumprecht et al., 1995; Bucci et al., 1996; LaBorde et al., 1997). In all cases where toxicity was evident it involved liver and/or kidney or homologous organs.

7.3 Mechanisms of toxicity — mode of action

Several biochemical modes of action have been proposed to explain all or some of the fumonisin-induced animal diseases. Two of these invoke disruption of lipid metabolism as initial site of action. There are also several studies that hypothesize fumonisin-induced changes in key enzymes involved in cell cycle regulation, differentiation and/or apoptosis as initial or secondary sites of action.

7.3.1 Disruption of sphingolipid metabolism

The structural similarity between sphinganine and FB_1 led Wang et al. (1991) to hypothesize that the mechanism of action of this mycotoxin might be via disruption of sphingolipid metabolism or a function of sphingolipids. At the moment, there are considerable data supporting the hypothesis that fumonisin-induced disruption of sphingolipid metabolism is an important event in the cascade of events leading to altered cell growth, differentiation and cell injury observed both *in vitro* and *in vivo*.

7.3.1.1 Sphingolipids and their metabolism

The pathways of biosynthesis and turnover (Fig. 1) have not been as well studied in sphingolipids as in other lipid classes. In order to understand how disruption of sphingolipid metabolism might contribute to the farm animal and laboratory animal diseases associated with consumption of fumonisins, it is necessary to understand how sphingolipids are biosynthesized. Eukaryotic cells synthesize a diverse array (over 400 distinct molecules) of sphingolipids which serve as important structural molecules in membranes and as regulators of many cell functions (Bell et al., 1993). While sphingolipids have also been found in procaryotes (Karlsson, 1970), their biosynthesis and role in cellular regulation is poorly understood.

Typically, *de novo* sphingolipid biosynthesis proceeds via the reactions described below (Merrill & Jones, 1990; Sweeley, 1991; Bell et al., 1993). The first is the condensation of serine with palmitoyl-CoA by serine palmitoyltransferase, a pyridoxal 5′-phosphate-dependent enzyme, and the resulting 3-ketosphinganine is reduced to sphinganine using NADPH. Sphinganine is acylated to dihydroceramides (also called *N*-acylsphinganines) by ceramide synthase using various fatty acyl-CoAs. Headgroups (e.g., phosphorylcholine, glucose, etc.) are subsequently added to the 1-hydroxyl group. The 4,5-trans-double bond of the sphingosine backbone is added after acylation of the amino group of sphinganine by the enzyme dihydroceramide desaturase (Michel et al., 1997). Both dihydroceramide and dihydrosphingomyelin are substrates for the enzyme. Thus, free sphingosine is not an intermediate of *de novo* sphingolipid biosynthesis (Merrill, 1991; Rother et al., 1992). Sphingolipid turnover is thought to involve the hydrolysis of complex sphingolipids to ceramides, then to sphingosine. Sphingosine is either reacylated or phosphorylated and cleaved to a fatty aldehyde and ethanolamine phosphate. The fatty aldehyde and ethanolamine phosphate can be redirected into the biosynthesis of glycerophospholipids and other fats (Van Veldhoven & Mannaerts, 1993).

In animal cells the initial steps from the condensation of serine and palmitoyl-CoA to the formation of ceramide take place in the endoplasmic reticulum. Subsequent processing of ceramide into glycosphingolipids and sphingomyelin takes place in the endoplasmic reticulum and Golgi apparatus. Degradation of complex sphingolipids occurs in the lysosomes, endosomes and the plasma membrane with degradation of free sphingoid bases occurring in the cytosol. For reviews of sphingolipid metabolism, see Merrill & Jones (1990), Merrill (1991), Sweeley (1991) and the volumes edited by Bell et al. (1993).

7.3.1.2 *Fumonisin-induced disruption of sphingolipid metabolism* in vitro

The term "fumonisin disruption of sphingolipid metabolism" includes inhibition of sphingosine and ceramide biosynthesis, depletion of more complex sphingolipids, increase in free sphinganine, decrease in reacylation of sphingosine derived from complex sphingolipid turnover and degradation of dietary sphingolipids, increase in sphingoid base degradation products (i.e. sphingosine (sphinganine) 1-phosphate, ethanolamine phosphate and fatty aldehydes), and increase in lipid products derived from the increase in the sphingoid base degradation products. FB_1 is now widely used to reveal the function of sphingolipids and sphingolipid metabolism in cells (Merrill et al., 1996a).

Fig. 1a

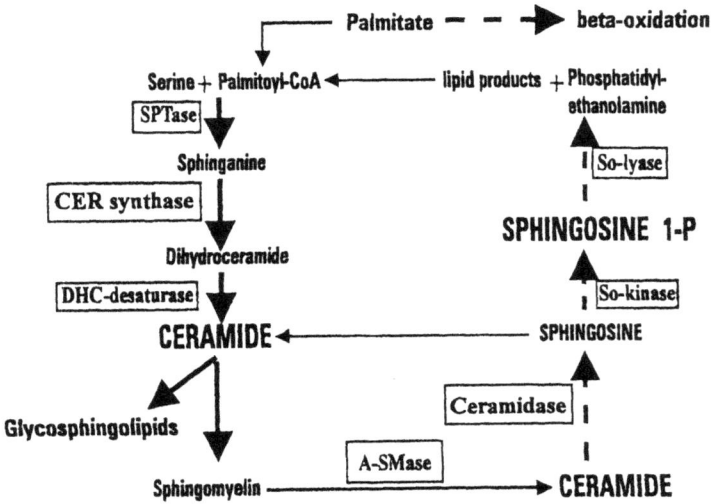

Fig.1b

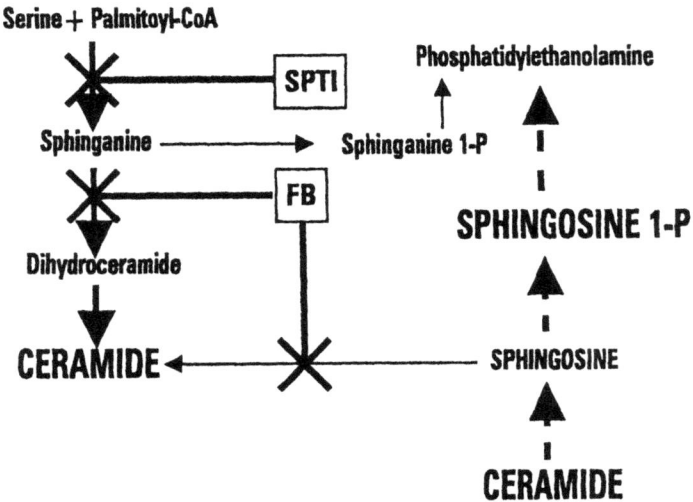

Fig. 1a. The pathway of *de novo* sphingolipid biosynthesis and turnover in a mammalian cell. Large solid arrows indicate the enzymatic steps leading to biosynthesis of ceramide, a known effector of cell death, and large broken arrows show the enzymatic steps leading to the production of sphingosine 1-phosphate, an effector of cell survival. Also shown is the proposed role of mitochondrial perturbations triggering a redirection of palmitate from beta-oxidation into the *de novo* pathway, resulting in increased biosynthesis of ceramide under conditions of oxidative stress.

Fig. 1b. The sites of action of serine palmitoyltransferase (SPTase) inhibitors (SPTI) such as ISP-I = myriocin = thermozymozydin, and ceramide synthase (CER synthase) inhibitors (FB) such as fumonisins. The block on the ceramide synthase responsible for reacylation of sphingosine results in an increase in free sphingosine and possibly sphingosine 1-phosphate. However, it has been reported that unlike sphingosine 1-phosphate, sphinganine 1-phosphate does not exert a marked cytoprotective effect, but does bind to and signal via the G protein-coupled receptor encoded by endothelial differentiation gene 1 (EDG1) (Spiegel, 1999).

In both Fig 1a and Fig 1b, the biochemicals shown in all capital letters are those known or suspected to be lipid secondary messengers. Also shown in Fig 1a is the generation of ceramide by ligand-induced sphingomyelin hydrolysis.

Abbreviations: DHC-desaturase (dihydroceramide desaturase), A-SMase (acidic sphingomyelinase), So-kinase and -lyase (sphingosine kinase and lyase).

Fumonisins potently inhibit the acylation of sphinganine and sphingosine (Wang et al., 1991; Yoo et al., 1992; Merrill et al., 1993c). In primary rat hepatocytes, the IC_{50} for inhibition of serine incorporation into sphingosine is approximately 0.1 μM for FB_1 and FB_2 (Wang et al., 1991; Merrill et al., 1993c). In P388 murine macrophages, the IC_{50} is less than 0.5 μM (Balsinde et al., 1997). In cultured pig renal cells (LLC-PK_1) the IC_{50} for inhibition of *de novo* sphingosine biosynthesis is approximately 20 μM FB_1 (Yoo et al., 1992). The basis for this difference in sensitivity is unknown.

Inhibition of sphinganine (sphingosine) *N*-acyltransferase (ceramide synthase) in cells leads to a concentration-dependent reduction in total complex sphingolipids, including sphingomyelin and glycosphingolipids (Wang et al., 1991; Merrill et al., 1993b; Yoo et al., 1996). In LLC-PK_1 cells, the decrease in complex sphingolipids does not become apparent until 24 to 48 h after cells have been exposed to FB_1 but before inhibition of cell growth and increased cell death (Yoo et al., 1996). In microsomal preparations from cultured mouse cerebellar neurons, inhibition of ceramide synthase was competitive with respect to both the long-chain (sphingoid) base and fatty acyl-CoA

(Merrill et al., 1993b). This observation suggests that ceramide synthase recognizes both the amino group (sphingoid binding domain) and the tricarballylic acid side-chains (fatty acyl-CoA domain) of FB_1 (Merrill et al., 1996b). The reduced inhibition by the hydrolysed derivatives of the fumonisin B series (Norred et al., 1997; van der Westhuizen et al., 1998) supports this hypothesis. Ceramide synthase has recently been shown to acylate hydrolysed FB_1, AP_1, to form N-palmitoyl-AP_1. The product was also found to be an inhibitor of ceramide synthase and to be 10 times more toxic than FB_1 in a human colonic cell line, HT29 (Humpf et al., 1998). Sphingomyelin biosynthesis is approximately 10-fold more sensitive to inhibition by FB_1 than glycosphingolipids. This is true for other inhibitors of ceramide synthesis, such as β-fluoroalanine (Medlock & Merrill, 1988; Merrill et al.,1993b).

The complete inhibition of ceramide synthase by fumonisins causes the intracellular sphinganine concentration to increase rapidly (Wang et al., 1991; Yoo et al., 1992). The amount of free sphinganine that accumulates in cells is a function of several factors. These include the extent of inhibition of ceramide synthase, the concentration of essential precursors (serine, palmitoyl-CoA), the growth rate of the cells, the rate of sphinganine degradation, and the rate of elimination from the cells. For example, in rat primary hepatocytes treated with a concentration of 1 μM FB_1, which inhibits serine incorporation into sphingosine by > 90%, there was a significant increase in free sphinganine after 1 h, which increased to 110-fold over controls after 4 days (Wang et al., 1991). In LLC-PK₁ cells approximately 50-fold and 128-fold increases were measured after exposure to FB_1 (35 μM, 50 to 60% inhibition) for 6 h and 24 h, respectively (Yoo et al., 1992). Proliferating LLC-PK₁ cells accumulate much higher levels of free sphingoid bases than confluent monolayers and cytotoxicity is only observed in proliferating cells (Yoo et al., 1996).

While the level of free sphingosine does not increase in primary rat hepatocytes (Wang et al., 1991; Gelderblom et al., 1995), it does in LLC-PK₁ cells, presumably due to inhibition of reacylation of sphingosine derived from sphingolipid turnover or from the growth medium. When cells begin to die, sphingosine levels will increase due to the breakdown of membrane lipids. However, in LLC-PK₁ cells the increase in free sphingosine occurs before any evidence of increased cell death or inhibition of cell proliferation (Yoo et al., 1992). Nonetheless, approximately 95% of the increase in the levels of free sphingoid bases in LLC-PK$_1$ cells was found to be due to the increase in free sphinganine level (Yoo et al., 1992). The fate of the accumulated sphinganine is unclear. While sphingosine has little difficulty in crossing cell membranes (Hannun et al., 1991), the half-life of sphinganine inside LLC-PK$_1$ cells is much longer than the half-life of FB$_1$ in LLC-PK$_1$ cells (Riley et al., 1998), which suggests either that the inhibition of sphinganine N-acyltransferase is persistent, that sphinganine does not easily diffuse out of cells, or that sphinganine degradation is slow relative to its biosynthesis. In urine from rats fed FB$_1$, > 95% of the free sphinganine is recovered in dead cells which have apparently sloughed into the urine (Riley et al., 1994a). Thus, in urine sphinganine is tightly associated with the cells.

In rat hepatocytes, a portion of the accumulated sphinganine is metabolized to sphinganine 1-phosphate and then cleaved into a fatty aldehyde and ethanolamine phosphate (Merrill et al., 1993c), both of which can be redirected into other biosynthetic pathways. The enzyme responsible for hydrolysis of sphingosine (sphinganine) 1-phosphate is sphingosine-phosphate lyase (Van Veldhoven & Mannaerts, 1993). In J774 cells exogenous sphinganine has been shown to be initially accumulated and then rapidly metabolized (Smith & Merrill, 1995). About one-third of the ethanolamine in phosphatidylethanolamine is derived from long-chain base catabolism when fumonisin is added (Smith & Merrill, 1995). Similar findings have been reported using fumonisin-treated Chinese hamster ovary cells (Badiani et al., 1996), confirming that, in some cell types, accumulated free sphingoid bases are rapidly metabolized into ethanolamine and fatty aldehydes. *In vitro* and *in vivo*, rat liver lipid composition is markedly altered by FB$_1$ (Wang et al., 1991; Gelderblom et al., 1997). In addition to changes in free sphingoid bases, more complex sphingolipids (ceramides, etc.) and phosphatidylethanolamine (Wang et al., 1991; Merrill et al., 1993c), there are many changes in the fatty acid composition of liver phospholipids (Gelderblom et al., 1997). The ability of cells to rapidly metabolize bioactive sphingoid bases into other products may protect cells from the toxicity associated with accumulation of free sphingoid bases or ceramide (Spiegel, 1999). Chronically disrupted sphingolipid metabolism leads to imbalances in phosphoglycerolipid and fatty acid metabolism. Because the accumulation of specific end-products and intermediates is dependent upon the balance between anabolic and catabolic processes, it is conceivable that changes in concentration of specific end-products could occur with no change in the steady-state concentration of free sphingoid bases.

7.3.1.3 *Fumonisin disruption of sphingolipid metabolism* in vivo

a) *Equids*

Typically free sphingosine and sphinganine are present in normal tissues and cells in trace amounts (Merrill et al., 1988; Riley et al., 1994c). This is to be expected since free sphinganine is a metabolic intermediate in the sphingolipid biosynthetic pathway, and free sphingosine is generated primarily as a consequence of sphingolipid turnover or degradation (Merrill, 1991; Rother et al., 1992; Michel et al., 1997). The first *in vivo* study to test if dietary fumonisins could change free sphingoid base concentration used serum obtained from ponies fed diets containing maize screenings naturally contaminated with fumonisins (primarily FB₁) (Wang et al., 1992). Upon consumption of diets containing fumonisin, all of the ponies exhibited large increases in sphinganine, although the magnitude of the changes varied among the animals. The elevation in serum sphinganine is reversible and the increase in free sphinganine and the Sa/So ratio occurred before increases in serum transaminase activity and clinical signs of ELEM (Wang et al., 1992). For example, a pony was given feed (corn screenings) containing 15 to 22 mg FB₁/kg. The Sa/So ratio increase by day 182 was followed by an increase in serum biochemical indices of cellular injury by day 223 (Wang et al., 1992). The pony died of ELEM on day 241.

In another study, horses were fed diets containing *F. proliferatum* (M 6290 and M 6104) culture material, which contained primarily either FB₂ (75 mg/kg) or FB₃ (75 mg/kg) (Ross et al., 1994). Analysis of serum and tissues from these horses showed a qualitatively similar response to horses fed FB₁ (Wang et al. 1992), but the magnitude of the increase in free sphinganine in serum, liver and kidney was much less in the horses fed FB₃ diets (Riley et al., 1997). While two of the three ponies fed FB₂ diets developed ELEM, the three ponies fed FB₃ diets showed no clinical signs of ELEM or evidence of liver damage after 60 days (Ross et al., 1994).

There was also a reduction in the amount of complex sphingolipids in the serum, liver and kidney of ponies fed diets containing fumonisins (Wang et al., 1992; Riley et al., 1997), as would be expected if the elevation in sphinganine was due to inhibition of *de novo* sphingolipid biosynthesis. In ponies fed diets containing FB_2 or FB_3 (75 mg/kg), the levels of complex sphingolipids in the liver were reduced by 88% and 72%, respectively, and the kidney was similarly affected (Riley et al., 1997). In contrast to ponies fed FB_2 diets, ponies fed FB_3 diets exhibited no liver or kidney pathology (Ross et al., 1994). Thus, prolonged exposure to fumonisins can result in a marked depletion of the complex sphingolipids in liver with no evidence of liver pathology.

b) *Pigs*

Similar results were obtained for pigs, confirming that there was a dose–response relationship between the ratio of free sphinganine to free sphingosine in serum and tissues and the amount of fumonisin-contaminated feed consumed (Riley et al., 1993). Pigs were fed diets formulated from naturally contaminated maize screenings at 0 (< 1), 5, 23, 39, 101 and 175 mg/kg total fumonisins (FB_1 plus FB_2). The results showed that the Sa/So ratio was significantly elevated in the liver, lung and kidney from pigs consuming feeds containing $\geq$ 23 mg/kg fumonisins. Liver injury was observed at fumonisin levels $\geq$ 23 mg/kg. However, injury to the kidney was not observed at any dose even though it contained equal or greater amounts of free sphingoid bases. In lung tissue, free sphingoid base content was elevated at doses $\geq$ 23 mg/kg, but lung lesions were only observed in pigs fed the diet containing 175 mg/kg. Smith et al. (1996a) showed that in pigs fed fumonisins significant effects on cardiovascular function were associated with significant increases in free sphingoid bases in heart tissue. Subsequent studies found that damage to pig alveolar endothelial cells, *in vivo*, was preceded by accumulation of free sphingoid bases in lung tissue (Gumprecht et al., 1998).

Elevation of the Sa/So ratio in pig serum paralleled the increase in tissues (Riley et al., 1993). This finding supported the earlier hypothesis (Wang et al., 1992) that the elevated ratio in serum was due to the movement of free sphinganine (accumulating as a result of inhibition of sphinganine *N*-acyltransferase) from tissues into the blood. Statistically significant increases in the serum ratio were observed at feed concentrations as low as 5 mg/kg total fumonisins (after 14 days) and in pigs (at higher concentrations) in which other serum biochemistry parameters were not changed and in which there were no observable gross or microscopic lesions in liver, lung or kidney. Thus, the increase in the Sa/So ratio was an earlier and more sensitive indicator of fumonisin exposure than the development of lesions in liver or lung in pigs detectable by light microscopy. Nonetheless, the increases in free sphinganine in tissues and serum closely paralleled the dose-dependent increases in other biochemical parameters measured at 14 days (Motelin et al., 1994).

It has been proposed that the ratio of free sphinganine to free sphingosine and the presence of elevated levels of free sphinganine in serum, urine and tissue be used as indicators for consumption of fumonisins by farm animals (Riley et al., 1994c). However, a subsequent study with pigs found altered growth at doses of FB₁ that did not cause an increase in free sphinganine (Rotter et al., 1996). Thus, in pigs, elevation of free sphinganine appears to occur at dosages that are greater than those that cause subtle changes in performance but lower than those that are toxic.

c) *Poultry and other commercially important animals*

Chickens fed diets supplemented with *F. verticillioides* culture materials (Weibking et al., 1993a, 1995) or pure FB₁ (Henry, 1993) exhibited elevated sphinganine levels and elevated ratios in tissues and serum. Similar findings have been made in the rabbit (Gumprecht, et al., 1995; LaBorde et al., 1997), catfish (Goel et al., 1994), mink (Restum et al., 1995; Morgan et al., 1997) and trout (Meredith et al., 1998).

d) *Laboratory animals*

In short-term studies with rats, rabbits and mice, disruption of sphingolipid metabolism, as shown by statistically significant increases in free sphinganine concentration, occurs at or below the fumonisin dosages than cause liver or kidney lesions (Riley et al., 1994a; Martinova & Merrill, 1995; LaBorde et al., 1997; de Nijs, 1998; Tsunoda et al., 1998; Voss et al., 1998). In rats (Sprague-Dawley, RIVM:WU) and mice (BALB/c) dosed with fumonisins, the increase in free sphinganine concentration in the kidney and/or liver is closely correlated with the extent and severity of lesions (Riley et al., 1994a; de Nijs, 1998; Tsunoda et al., 1998; Voss et al., 1998). In two separate 21-day feeding studies (Fischer-344 rats), liver free sphinganine level was increased, although not significantly, at the lowest FB_1 dose (50 mg/kg diet) that had liver cancer-promoting potential (Gelderblom et al., 1996c, 1997). In rats and rabbits, the concentration of fumonisin that causes nephrotoxicity and an increase in kidney free sphinganine concentration is lower than the fumonisin dose that causes hepatotoxicity (Voss et al., 1993, 1996b, 1998; Gumprecht et al., 1995; LaBorde et al., 1997; de Nijs, 1998). For example, in Sprague-Dawley rats significant elevation of free sphinganine levels and hepatosis were observed at ≥ 15 mg/kg ≤ 50 mg/kg dietary FB_1 (Riley et al., 1994a), whereas the NOEL for nephrosis in male Sprague-Dawley rats is 9 mg/kg (Voss et al., 1995) and significant increases in kidney free sphinganine have been detected in rats fed AIN-76 diets containing 1 mg FB_1/kg (Wang et al., 1999). In male RIVM:WU rats, the liver free sphinganine level was significantly elevated at $> 0.19 \leq 0.75$ mg FB_1/kg body weight (equivalent to 1.9 and 7.5 mg/kg dietary FB_1) in the absence of any evidence of hepatosis (de Nijs, 1998) and the NOEL for tubular cell death and significant increases in kidney free sphinganine was ≤ 0.19 mg FB_1/kg body weight, which was equivalent to 1.9 mg/kg in feed (de Nijs, 1998).

In the US National Toxicology Program, long-term feeding study with Fischer-344/N Nctr BR rats, pure FB_1 induced an increase in the Sa/So ratio in kidney tissue, which correlated with increased non-neoplastic and neoplastic lesions (US NTP, 1999). In $B6C3F_1$/Nctr BR female mouse liver, free sphinganine and the Sa/So ratio were increased after 3 and 9 weeks at 50 and 80 mg FB_1/kg diet, which were the same doses that induced liver adenoma and carcinoma (US NTP, 1999). Livers taken from rats (BD IX) in a long-term study (2 years) that were fed diets containing 10 and 25 mg/kg FB_1 did not show significant changes in liver free sphingoid bases, although the mean concentration of free sphinganine and free sphingosine in the liver of rats fed 25 mg FB_1/kg diet was 8- and 3-fold, respectively, higher than control values (Gelderblom et al., 1997).

In rats, rabbits and vervet monkeys, increases in free sphinganine concentration have been detected in the urine of animals fed fumonisin-containing diets (Riley et al., 1994a; Castegnaro et al., 1996; Shephard et al., 1996b; LaBorde et al., 1997; Merrill et al., 1997b; Solfrizzo et al., 1997a,b; Wang et al., 1999). Accumulation of free sphinganine in rat urine (associated with accumulation of dead cells) closely reflected the changes which occurred in the kidney (Riley et al., 1994a). Analysis of urine from rats fed commercially available chows showed a statistically significant correlation between the free sphinganine to free sphingosine ratio in urine and the fumonisin concentration in the chows. The FB_1 concentration in the chows ranged from undetected to 3.3 mg/kg (Merrill et al., 1997b). Feeding studies with pure FB_1 in AIN-76 diets indicate that the no-observed-effect level (NOEL) for elevation of urinary free sphinganine level in Sprague-Dawley rats is > 1 mg/kg diet ≤ 5 mg/kg diet (Wang et al., 1999). In other studies, rats fed a diet containing mixture of fumonisins (from culture material) for 13 days showed a NOEL of between 1 and 2 mg/kg diet (Solfrizzo et al., 1997b).

In rats fed an AIN-76 diet containing 10 mg FB_1/kg for 10 days and then put on a control diet, the urinary sphinganine concentration returned to control levels in 10 days. However, if the diet contained 1 mg FB_1/kg, the urinary sphinganine concentration remained markedly elevated for at least 10 days after changing the feed (Wang et al., 1999). Thus, in this study, once elevated by feeding toxic levels of FB_1, apparently non-toxic concentrations kept the free sphinganine concentration significantly elevated to concentrations that were equivalent to those of the nephrotoxic fumonisin dosage. This result is, however, in contrast with the findings of Solfrizzo et al. (1997b), showing that the elevated levels of sphingoid bases after exposure to relatively high levels of fumonisins (7–15 mg/kg for 13 days) return to their original values when rats are exposed to a low-fumonisin diet (1 mg/kg or less) for a period of time that is directly dependent on the previous level of exposure, in terms of dose and time (Solfrizzo et al., 1997b).

7.3.1.4 Tissue and species specificity

The tissue specificity and the severity of the pathology in rats (Sprague-Dawley, Fischer-344, Wistar, RIVM:WV) and mice (BALB/c) seem to correlate well with the disruption of sphingolipid biosynthesis (Riley et al., 1994a; Tsunoda et al., 1998; Voss et al.,

1998). This is not the case in pigs and horses, where the kidney appears to be equally or more sensitive than the liver with regards to the fumonisin-induced increase in free sphinganine (Riley et al., 1993, 1997). There is significant liver pathology in horses and pigs (Ross et al., 1994; Haschek et al., 1996), with little evidence of kidney damage (Harvey et al., 1996).

It has been suggested that these differences in tissue and species specificity may be due to differing susceptibility to the adverse cellular effects of disrupted sphingolipid metabolism (Voss et al., 1996b). For example, liver and kidney may have different abilities to metabolize or eliminate free sphinganine or to compensate for depletion of complex sphingolipids. In addition, fumonisin, free sphingoid bases or their metabolites, in serum may affect the function of the vasculature and thus indirectly affect tissues that are not directly affected by fumonisin inhibition of ceramide synthase (Ramasamy et al., 1995; Smith et al., 1996b). For example, the correlation between the fumonisin-induced increase in serum free-sphinganine (Wang et al., 1992) and the onset of ELEM could be explained if the vascular function in horse brain was altered due to elevated serum free sphinganine. In pigs, it has been hypothesized that cardiovascular dysfunction, subsequent to increased free sphingoid base concentration in the heart, is the cause of PPE (Smith et al., 1996b).

Riley et al. (1996) have recommended that detection of high concentrations of free sphinganine in urine, serum or tissues should be viewed as a clinical tool to be developed and used in conjunction with other clinical tools in situations where animal toxicity resulting from exposure to fumonisins is suspected. Changes in sphingolipid profiles in serum and urine in vervet monkeys fed fumonisin-containing diets have been reported (Shephard et al., 1996b). Whether human exposure to fumonisins in maize and maize products will result in increased free sphinganine concentration in tissues, urine or serum is not known. However, free sphingoid bases can be detected in human urine (Castegnaro et al., 1996; Solfrizzo et al., 1997a).

7.3.1.5 *Fumonisin-induced sphingolipid alterations: effects on growth, differentiation and cell death*

There are many ways that disruption of sphingolipid metabolism could account for the cell damage caused by fumonisins. In order to fully understand the possibilities, it is necessary to consider the multitude of functions of complex sphingolipids (Bell et al., 1993), the potent bioactivity of sphinganine and its metabolites (Merrill et al., 1993a), and the parallel or branch metabolic pathways that can be affected by disruption of sphingolipid metabolism (Riley et al., 1996; Merrill et al., 1997a,b). Since the steady-state concentration of many biologically active lipid intermediates and end-products could be altered, there are also many potential molecular sites that could be affected by fumonisin-induced disruption of sphingolipid metabolism. Thus, it can be expected that there will also be a diversity of alterations in cellular regulation.

The earliest effect of fumonisin on sphingolipid metabolism *in vitro* is the decrease in serine incorporation into ceramide, followed by an increase in free sphinganine concentration (Yoo et al., 1992). There is also a concentration-dependent decrease in more complex sphingolipids (Yoo et al., 1996). Because long-chain (sphingoid) bases are growth inhibitory, cytotoxic and induce apoptosis under some conditions (Merrill, 1983; Stevens et al., 1990; Nakamura et al., 1996; Sweeney et al., 1996; Yoo et al., 1996), and are growth stimulating under certain conditions (Zhang et al., 1990, 1991; Schroeder et al., 1994), the accumulation of sphinganine (and sometimes sphingosine) might account for these same effects of fumonisins. Yoo et al. (1992) have shown that in the renal epithelial cell line (LLC-PK₁ cells) there is a concentration-dependent association between the inhibition of sphingolipid biosynthesis by FB₁ and growth inhibition and cell death. After 24 h of exposure to FB₁ many cells began to develop a fibroblast-like appearance, with loss of cell-cell contact and an elongated, spindle shape. If fumonisin was removed, the cells that survived resumed growth and had a normal epithelial morphology. Addition of exogenous sphinganine induces cell death at intracellular concentrations that are similar to those induced by FB₁ (Yoo et al., 1996).

The two most likely explanations for the increased cell death after inhibition of sphingolipid biosynthesis by fumonisins are: (1) that the free sphinganine (or a sphinganine degradation product) is growth inhibitory and cytotoxic for the cells, as has been seen in many other systems (Stevens et al., 1990; Hannun et al., 1991; Sweeney et al., 1996); and (2) that more complex sphingolipids are required for cell survival and growth, as has been proven with mutants lacking serine palmitoyltransferase (Hanada et al., 1990, 1992) and in studies with specific inhibitors of glycosphingolipid biosynthesis (Radin, 1994; Nakamura et al., 1996). β-Chloroalanine, a non-specific serine palmitoyltransferase inhibitor, in the presence of FB_1 reduced the intracellular concentration of free sphinganine and also reduced the inhibition of cell growth (50 to 60%) and the extent of cell death (50 to 60%) (Yoo et al., 1996). More recent studies with LLC-PK_1 cells indicate that fumonisin inhibition of cell proliferation and increased cell death (apoptosis) are prevented by > 90% using the specific serine palmitoyltransferase inhibitor, myriocin (ISP-1) (Riley et al., 1999). Similar results have been obtained with HT29 cells, a human colonic cell line (Schmelz et al., 1998). However, in the LLC-PK_1 cells, the morphological changes, such as decreased cell-cell contact and increased fibroblast-like appearance, are not reversed. In primary human keratinocytes, both β-chloroalanine and *N*-acetylsphingosine partially protected against FB_1-induced apoptosis (Tolleson et al., 1999). However, both exogenous sphinganine and *N*-acetylsphingosine alone induced apoptosis in these same cells (Tolleson et al., 1999). Thus, in cultured cells sphingolipid-dependent mechanisms for inducing apoptosis include accumulation of excess ceramide or sphingoid bases, or depletion of ceramide, or more complex sphingolipids.

In addition to the cell types described above, apoptosis in response to exposure to FB_1 *in vitro* has been reported using turkey lymphocytes (Dombrink-Kurtzman et al., 1994a,b), human fibroblasts, oesophageal epithelial cells and hepatoma cells (Tolleson et al., 1996b), and CV-1 monkey kidney cells (Wang et al., 1996).

The adverse effects of fumonisin-induced depletion of more complex sphingolipids have been demonstrated in numerous other studies. For example, in hippocampal neurons, FB_1 inhibition of complex sphingolipid biosynthesis was correlated with decreased axonal growth (Harel & Futerman, 1993). The FB_1 inhibition of axonal growth could be reversed by addition of ceramide with FB_1 (Harel & Futerman, 1993; Schwarz et al., 1995). The ability of growth factors to stimulate axonal cell growth is dependent on sphingolipid biosynthesis (Boldin & Futerman, 1997). In fibroblasts (Swiss 3T3 cells), fumonisin-induced morphological changes could be reversed by ganglioside GM_1. However, GM_1 did not prevent the inhibition of cell

71

proliferation (Meivar-Levy et al., 1997). FB_1 and/or myriocin (ISP-1) inhibition of glycosphingolipid biosynthesis disrupts cell substratum adhesion in mouse melanoma cells (Hidari et al., 1996). FB_1 has also been shown to alter the manner in which glycosyl phosphatidylinositol-anchored proteins, such as the folate receptor, are organized and function in membranes (Hanada et al., 1993; Stevens & Tang, 1997). FB_1 inhibition of galactosylceramide biosynthesis has been shown to disrupt the assembly and disassembly of cytoskeletal proteins responsible for lipid transport and maintenance of the subcellular architecture in SW13 cells (derived from a human adrenal carcinoma) (Gillard et al., 1996). Thus, there is no doubt that the loss of complex sphingolipids also plays a role in the abnormal behavior and altered morphology of fumonisin-treated cells.

Currently Swiss 3T3 cells are the only type of cell that respond to fumonisins with increased DNA synthesis (Schroeder et al., 1994). It was proposed that this *in vitro* model would be useful for understanding if, within the complex *in vivo* milieu of cells in the liver, there might exist cells that could be inappropriately selected to enter the cell cycle. Defects in cell cycle control have been shown to promote genomic instability and progression to malignancy (Hartwell & Kastan, 1994). Incubation of Swiss 3T3 cells with DL-erythro-sphinganine caused an increase in [³H]thymidine incorporation into DNA. Addition of FB_1 to the cells elevated sphinganine and induced a comparable increase in [³H]thymidine incorporation into DNA. These findings associated an accumulation of sphinganine with the induction of DNA synthesis by FB_1 but did not prove that they were causally linked. However, this was proven using an inhibitor of serine palmitoyltransferase in combination with FB_1. Reduction in cellular sphinganine when β-fluoro-L-alanine was added to Swiss 3T3 cells, demonstrated that this reduction in sphinganine completely removed the insulin-dependent stimulation of [³H]thymidine incorporation into DNA by FB_1. Therefore, the formation of sphinganine is required for stimulation of DNA synthesis by fumonisins in Swiss 3T3 cells (Schroeder et al., 1994).

Fumonisin inhibition of ceramide synthesis can deregulate many normal cell functions including non-accidental programmed cell death. Some of the processes have been shown to be modulated by fumonisin inhibition of ceramide synthase *in vitro* (Table 4).

Table 4. Examples of cell functions modulated by *de novo* ceramide biosynthesis as shown by inhibition of the process by fumonisin-treatment

- sphingosine-induced germinal vesicle breakdown and *Xenopus* oocyte maturation (Strum et al., 1995)

- daunorubicin-activated apoptosis in P388, U937 and chicken granulosa cells (Bose et al., 1995; Witty et al., 1996)

- chemotherapeutic agent (CPT-11)-induced interleukin 1-beta converting enzyme (ICE) cascade-dependent apoptosis in 4B1 (L929) mouse fibroblasts (Suzuki et al., 1997)

- carnitine palmitoyltransferase inhibition-induced apoptosis in LyD9 mouse haematopoietic precursor cells (Paumen et al., 1997)

- lipopolysaccharide (LPS)/platelet activating factor (PAF) induced arachidonic acid release in murine $P388D_1$ macrophages (Balsinde et al., 1997)

- chemical hypoxia-induced cell death in $LLC-PK_1$ cells (Ueda et al., 1998)

- Fas-transduced-caspase-dependent T-cell proliferation (Sakata et al., 1998)

- fenretinide-induced poly-(ADP-ribose) polymerase (PARP) cleavage and apoptosis in HL-60 cells (DiPietrantonio et al., 1998)

- serum-stimulated retinoblastoma (Rb) protein dephosphorylation and cell cycle progression (Lee et al., 1998)

- multidrug resistance modulator-dependent cytotoxicity (Cabot et al., 1998, 1999)

- TNF-α/cycloheximide-induced endothelial cell death (Xu et al., 1998)

- 12-O-tetradecanoylphorbol-13-acetate (TPA)-induced apoptosis in prostate cancer cells (Garzotto et al., 1998)

- hexadecylphosphocholine-induced apoptosis in HaCaT cells (Wieder et al., 1998)

- fatty acid-induced nitric oxide synthase-dependent apoptosis in cultured rat prediabetic islets (Shimabukuro et al., 1998)

- ionizing radiation-induced DNA damaged and cell death in various cell types (Liao et al., 1999)

It is important to recognize that ceramide signaling is also mediated by sphingomyelin hydrolysis (Perry & Hannun, 1998) via enzymes that are not inhibited by fumonisin. When fumonisins are added to cells for the purpose of inhibiting *de novo* ceramide generation, there is also the potential for accumulation of free sphingoid bases and their downstream sphingoid base 1-phosphates. Thus, there is the potential for misinterpreting the results of experiments using fumonisins as an inhibitor of ceramide biosynthesis as was recently pointed out by Lemmer et al. (1998).

The ability of fumonisin inhibition of ceramide biosynthesis to protect cells is of interest since primary rat hepatocyte necrotic cell death has been shown to be mediated by ceramide-induced (but not dihydroceramide) mitochondrial dysfunction (Arora et al., 1997). The activity of the enzyme responsible for desaturation of the inactive dihydroceramide (dihydroceramide desaturase) is regulated by the intracellular redox state of the cell (Michel et al., 1997). Taken together, these findings suggest that ceramide metabolism is sensitive to oxidative stress and that fumonisin-inhibition of ceramide will modify apoptosis induced by mitochondrial damage or oxidative stress.

In primary rat hepatocytes and rat liver slices, large increases in free sphinganine occur at FB_1 concentrations ranging from 0.1 to 1 µM (Wang et al., 1992; Gelderblom et al., 1996b; Norred et al., 1996) that are 300- to 3000-fold less than those that are cytotoxic and 10- to 100-fold less than those that cause inhibition of epidermal growth factor-induced [³H]thymidine incorporation into DNA (Gelderblom et al., 1996b). There appears to be no relationship between fumonisin-induced increases in free sphinganine and fumonisin-induced inhibition of [³H]thymidine incorporation and cell death in primary rat hepatocytes (Gelderblom et al., 1996a,b) or the cytotoxicity in rat liver slices (Norred et al., 1997). At the moment there is no adequate explanation for the resistance of primary rat hepatocytes and rat liver slices to fumonisin-induced cytotoxicity or inhibition of cell proliferation. This is puzzling in light of the fact that other primary cell cultures, such as rabbit renal epithelial cells, are very sensitive to fumonisin-induced inhibition of cell growth (Counts et al., 1996) and that in liver, *in vivo*, the intracellular concentrations of FB_1 that cause hepatotoxicity are relatively low based on pharmacokinetic considerations (see previous sections).

7.3.1.6 *Sphingolipid-mediated cellular deregulation and fumonisin diseases*

Sphingolipids have been associated with many facets of cellular regulation (Merrill et al., 1993a; Bell et al., 1993; Ballou et al., 1996;

Merrill et al., 1997b; Kolesnick & Krönke, 1998; Perry & Hannun, 1998) that could contribute to or modify the expression of fumonisin-associated diseases (Table 5).

Table 5. Examples of cellular regulatory processes that have been shown to be modulated by sphingolipids and are known to be important in the control of normal cell growth, differentiation, apoptosis and immune response [a]

Sphingoid bases and their metabolites

- inhibition of protein kinase C

- activation of phospholipase D/inhibition of phosphatidic acid phosphatase

- activation of the epidermal growth factor (EGF) receptor kinase (probably via mitogen-activated protein kinase)

- control of intracellular calcium (seemingly via sphingosine 1-phosphate)

- control of plasma membrane potassium permeability in myocytes

- inhibition of DNA primase and increases in transcription factor AP-1, an early step in the growth of some cell types

Ceramide

- second messenger in cytokine signal transduction

- activates protein kinases, phosphatases and MAP kinases

- inhibits phospholipase D

More complex sphingolipids

- binding of cytoskeletal proteins

- participation in cell-cell communication and cell-substratum interactions

- protein transport, sorting and targeting

[a] For additional processes regulated by ceramide generated *de novo*, see Table 4 and the reviews cited in the text.

Chronic fumonisin disruption of sphingolipid metabolism has been hypothesized to be a contributing factor leading to cellular deregulation and organ toxicity (Merrill et al., 1993c; Riley et al., 1994b; Schroeder et al., 1994; Tolleson et al., 1996a,b, 1999). The consequences of disrupted sphingolipid metabolism *in vitro* are specific to cell type.

The evidence for fumonisin-induced disruption of sphingolipid metabolism in target tissues has been demonstrated repeatedly in many independent studies. Nonetheless, the precise mechanism by which disrupted sphingolipid metabolism contributes to the increased organ toxicity in rodents is unclear. The current understanding of the sphingolipid signalling pathways (Merrill et al., 1997a,b; Kolesnick & Krönke, 1998; Perry & Hannun, 1998; Spiegel, 1999) indicates that the balance between the intracellular concentration of sphingolipid effectors that protect cells from apoptosis (decreased ceramide, increased sphingosine 1-phosphate) and the effectors that induce apoptosis (increased ceramide, increased free sphingoid bases, increased fatty acids) will determine the observed cellular response (the critical set-points will be cell-type specific). Since the balance between the rates of apoptosis and proliferation are critical determinants in the process of tumorigenesis, in cells exposed to fumonisins, those cells sensitive to the proliferative effect of decreased ceramide and increased sphingosine 1-phosphate should be selected to survive and proliferate when the conditions under which the cells are exposed to fumonisins are such that increased intracellular free sphingoid base concentration is not growth inhibitory. Conversely, when the rate of increase in free sphingoid bases exceeds a cell's ability to convert sphing-anine/sphingosine to dihydroceramide/ceramide or their sphingoid base 1-phosphate, then free sphingoid bases will accumulate. In this case cells that are sensitive to sphingoid base-induced growth arrest will cease growing and insensitive cells will survive. Another condition that would promote increased apoptosis would be if the block on ceramide synthase was either reduced or ceramide synthesis was increased while free sphinganine levels were still high.

7.3.2 Altered fatty acid metabolism in liver

Gelderblom et al. (1995) reported that there was no relationship between the fumonisin-induced increase in free sphinganine and the mitoinhibitory effect or the cytotoxicity of FB₁ in primary rat hepatocytes. In addition it was found that free sphinganine concentration increased markedly even in primary rat hepatocytes that had not been exposed to FB_1. However, in cultured cells the simple act of changing the culture medium can result in a transient increase in free sphingoid bases (Smith & Merrill, 1995; Smith et al., 1997). In fumonisin-treated primary rat hepatocytes, the Sa/So ratio was maximally elevated at 1 μM FB_1, whereas cytotoxicity was observed at ≥ 250 μM FB_1 (Gelderblom et al., 1995, 1996b). Polyunsaturated fatty acids (PUFAs) were shown to accumulate at the cytotoxic doses (Gelderblom et al., 1996b).

In other studies, fumonisins were shown to create a multitude of changes in liver cholesterol, phospholipids, sphingoid bases and free fatty acid composition (Gelderblom et al., 1996a, 1997). In both the short-term and the long-term feeding studies, changes in fatty acid profiles indicated that FB_1 treatment altered the n-6 fatty acid metabolic pathway. In the long-term study (2 years), significant changes were observed in livers from rats fed 10 and 25 mg FB_1/kg diet (Gelderblom et al., 1997). These data suggested that the increase in the n-3 fatty acid content of liver could, through altered eicosanoid biosynthesis, modulate hepatocyte proliferation (Gelderblom et al., 1997). Recently, fumonisin treatment has been shown to increase the extent of lipid peroxidation in rat (Fischer-344) primary hepatocytes and liver *in vivo* in a concentration- and dose-dependent manner (Abel & Gelderblom, 1998). The increased susceptibility to lipid peroxidation may be a consequence of the other lipid changes described above.

7.3.3 Other biochemical changes

Numerous studies using fumonisins have found changes in cellular regulation and cell function (Table 6). Many of these effects could be relevant to the organ toxicity of fumonisins.

In conclusion, fumonisin-induced disruption of sphingolipid metabolism is observed both *in vitro* and *in vivo*. With the exception of primary rat hepatocytes, disruption of sphingolipid metabolism is closely correlated in both a time- and concentration-dependent manner with alterations in cell proliferation and increased cell death. *In vivo*, evidence for disruption of sphingolipid metabolism is closely

Table 6. Summary of studies using fumonisins that have found
changes in cellular regulation and cell function

- repression of expression of protein kinase C (PKC), AP-1-dependent transcription, stimulation of a cyclic AMP response element in CV-I African green monkey kidney cells (1–10 µM FB_1, 3 to 16 h) (Huang et al., 1995)

- decreased phorbol dibutyrate binding, increased cytosolic PKC activity, with both exogenous sphinganine and FB_1 to J774A.1 cells (Smith et al.,1997)

- inhibited phorbol dibutyrate binding in short-term incubations using crude cerebrocortical membrane preparation and both FB_1 and exogenous sphingosine (Yeung et al., 1996)

- activation of the mitogen-activated protein kinase (MAPK) in Swiss 3T3 cells with FB_1 (Wattenberg et al., 1996)

- inhibition of serine/threonine phosphatases (PP5, IC_{50} of 80 µM) in isolated enzyme preparations (Fukuda et al., 1996)

- over-expression of nuclear cyclin D1 and increased cyclin-dependent kinase 4 (CDK4) activity in rat livers obtained from a long-term feeding study and a 21-day feeding study with FB_1 (an abstract Ramljak et al., 1996)

- dephosphorylation of the retinoblastoma protein, repression of CDK2, and induction of two CDK inhibitors in CV-1 cells with FB_1 (Ciacci-Zanella et al., 1998)

- apoptosis inhibitor and protease inhibitor protection of CV-1 cells and primary human cells from FB_1-induced apoptosis (Ciacci-Zanella & Jones, 1999)

- increased TNF secretion in LPS-activated intraperitoneal macrophages from FB_1-treated mice (Dugyala et al., 1998)

- altered calcium homeostasis in frog (*Rana esculenta*) atrial muscle *in vitro* (Sauviat et al., 1991)

- glutathione depletion and lipid peroxidation in cultured cells (Azuka et al., 1993; Kang & Alexander, 1996; Sahu et al., 1998; Abado-Becognee et al., 1998; Yin et al., 1998) and *in vivo* (Lim et al., 1996; Abel & Gelderblom, 1998)

- stimulation of nitric oxide production (Rotter & Oh, 1996)

correlated with the onset and progression of *F. verticillioides*-associated diseases in pigs, horses, rabbits, mice and rats. However, disrupted sphingolipid metabolism is also observed in tissues that are not considered target organs (i.e., pig and horse kidney, pig heart, endothelial cells). Thus, fumonisin-induced disruption of sphingolipid metabolism could contribute both directly and indirectly to the diseases known to be caused by consumption of fumonisins. Fumonisins also affect other sites of cellular regulation that are apparently independent of the disruption of sphingolipid metabolism. However, disruption of various aspects of lipid metabolism and signal transduction pathways mediated by lipid second messengers appears to be an important aspect of all the various proposed mechanisms of action.

7.4 Factors modifying toxicity; toxicity of metabolites

Voss et al. (1996c) found that nixtamalization of *F. verticillioides* (MRC 826) culture material effectively eliminated FB_1, but the resulting material (containing hydrolysed FB_1) was less hepatotoxic but equally nephrotoxic when fed to rats. In a recent abstract, it was reported that pure FB_1 at 50.5 and 101 mg/kg diet (70 and 140 µmol/kg diet) was toxic to female $B6C3F_1$/Nctr mice when fed for 28 days, but FB_2, FB_3 and AP_1 were not hepatotoxic (Howard et al., 1999). These considerations are important in the evaluation of the potential of calcium hydroxide treatment for the detoxification of fumonisin-contaminated maize (Sydenham et al., 1995).

In naturally contaminated maize, the simultaneous occurrence of multiple fumonisins is likely. Therefore, the relative toxicity of the various fumonisins is of concern for hazard assessment. Gelderblom et al. (1993) found that the aminopentols (AP_1 and AP_2) of FB_1 and FB_2 did not act as cancer initiators in orally dosed male Fischer rats, although they were more toxic than the parent compounds in primary cultures of rat hepatocytes. AP_1 is less toxic than FB_1 (Flynn et al., 1994, 1997). The aminopentols of FB_1, FB_2 and FB_3 are also less effective inhibitors of sphinganine *N*-acyltransferase in rat primary hepatocytes and liver slices (Merrill et al., 1993c; Norred et al., 1997). In gestation day 9.5 rat embryos exposed *in vitro* to 0, 3, 10, 30, 100 or 300 µM AP_1 throughout the entire 45-h cultured period, significant increase in the incidence of abnormal embryos including neural tube defects (NTD) were observed at concentrations of 100 µM and above (Flynn et al., 1997). A recent study by Norred et al. (1997) found that the following mycotoxins had no effect on sphinganine levels in rat liver slices: aflatoxin B_1, cyclopiazonic acid, beauvericin, T-2 toxin, sterigmatocystin, luteoskyrin, verrucarin A, scirpentriol and zearalenone. Fumonisins FB_1, FB_2, FB_3, FB_4, FC_4, and hydrolysed FB_1, FB_2, FB_3 and Aal-toxin all caused significant elevation in free sphinganine (Norred et al., 1997). Fumonisin B_4, C_4 and AAl-toxin are the most effective inhibitors of sphinganine N-acyltransferase based on sphinganine accumulation in rat liver slices (Norred et al., 1997). However, their toxicity *in vivo* is unknown. Pure FB_3 was less effective than FB_1 or FB_2 in causing reduced weight gain in rats (Gelderblom et al., 1993) but FB_2 and FB_3 are equally effective inhibitors of sphinganine N-acyltranferase (Norred et al., 1997). A diet containing FB_3 was less effective than a diet containing FB_2 in inducing ELEM in ponies (Ross et al., 1994). The ability of diets containing FB_3 to disrupt sphingolipid metabolism *in vivo* was less than that of diets containing FB_2 (Riley et al., 1997). However, in primary rat hepatocytes, FB_3 was found to be more cytotoxic (Gelderblom et al., 1993). Acetylated FB_1 had no effect (relative to controls) on weight gain in rats nor did it have any cancer-initiating activity (Gelderblom et al., 1993). It does not cause sphinganine accumulation in liver slices (Norred et al., 1997) but did cause sphinganine accumulation in a study with primary rat hepatocytes (van der Westhuizen et al., 1998).

Because fumonisins occur naturally in combination with other fungal toxins (Chu & Li, 1994; Bottalico et al., 1995; Logrieco et al., 1995; Yamashita et al., 1995; Ueno et al., 1997), the possibility of toxic synergisms exists. There are numerous reports of additive effects but the dosages have been much greater than those known to occur naturally (Kubena et al., 1995a,b). Toxic synergisms have been reported in growing pigs fed diets containing culture material with FB_1 (50 mg/kg diet) and deoxynivalenol-contaminated wheat (4 mg/kg diet) (Harvey et al., 1996) and in a similar study with aflatoxin (2.5 mg/kg diet) and FB_1 (100 mg/kg diet) culture material (Harvey et al., 1995).

Several reports have been published indicating that no toxic synergism is observed in poultry. In turkeys fed a ration containing 200 mg FB_1 and 100 mg moniliformin/kg diet from 1 to 21 days of age, no additive or synergistic effects were observed (Bermudez et al., 1997). Female turkey poults (Nicholas Large Whites) from day of hatch to 3 week of age fed diets containing 300 mg FB_1, as well as 4 mg diacetoxyscirpenol or 3 mg ochratoxin A, exhibited additive or less than additive toxicity, but not toxic synergy (Kubena et al., 1997a).

In male broiler chicks from day of hatch to 19 or 21 days of age fed diets containing 300 mg FB_1, as well as 5 mg T-2 toxin/kg diet or 15 mg deoxynivalenol/kg diet from naturally contaminated wheat, toxic synergy was not observed for either of these toxin combinations (Kubena et al., 1997b).

Fumonisins inhibit the *in vitro* biosynthesis of glycosphingolipid receptors for cholera toxin and shiga-like toxins (Sandvig et al., 1996) and inhibit the accumulation of glycosphingolipids believed to be responsible for multidrug-resistance in certain cancer cells (Lavie et al. 1996). Glycosphingolipids are known to be receptors and adhesion sites for viruses, bacteria and fungi.

8. EFFECTS ON HUMANS

There has been one report of a disease outbreak characterized by abdominal pain, borborygmi and diarrhoea in India suspected to be associated with foodborne FB_1 (Bhat et al., 1997).

8.1 Transkei, South Africa

The only studies available were correlation studies, most of which indicated some relationship between oesophageal cancer rates and the occurrence of *F. verticillioides* (IARC, 1993).

A very high incidence of oesophageal cancer among the black population of the Transkei, South Africa, has been reported in several surveys (Jaskiewicz et al., 1987c; Makaula et al., 1996), some of which have been reviewed by IARC (1993). The incidence was higher in both sexes in the south (Butterworth and Kentani Districts) compared to the northern parts of the region (Bizana and Lusikisiki Districts).

Based on the performance of hybrids in small experimental plots, maize grows well in both areas (Rheeder et al., 1994). The sites are about 200 km apart and the northern area is about 500 m higher. Although some soil fertility factors were different between the high and low oesophageal cancer areas, there is no evidence that any nutrient was limiting at least for hybrid maize production (Rheeder et al., 1994). Farmers grow open-pollinated maize of varying genotypes passed on from farm to farm and season to season. Kernel types include large flour-maize kernels as well as dent and flint-type white, yellow and blue kernels. Both areas depend on home-grown maize for around 50–100% of the year's supply, the remaining being purchased from commercial sources or imports.

Maize porridge is the staple diet (up to 100% of calories). Adults also consume beer deliberately made from mouldy maize selected by the housewife from the harvest. This maize has been found to contain up to 118 mg/kg fumonisins (Rheeder et al., 1992). Based on experiments conducted on beer made from wort containing added FB_1 (Scott et al., 1995), such beers could contain fumonisin concentrations of 30 mg/litre beer.

Contamination of home-grown maize in Transkei by a number of toxigenic *Fusarium* species, particularly *F. verticillioides,* has been observed since the early 1970s (Marasas et al., 1979a, 1981, 1988b, 1993; Marasas, 1993, 1994, 1995, 1996, 1997). Another *Fusarium* species associated with maize in Transkei is *F. graminearum,* and the mycotoxins deoxynivalenol and zearalenone produced by this fungus occur in home-grown maize intended for human consumption (Marasas et al., 1979b). However, the occurrence of *F. graminearum* in maize kernels was found to be greater in low-risk than in high-risk areas for oesophageal cancer in later studies (Marasas et al., 1981, 1988b), and Sydenham et al. (1990b) confirmed that deoxynivalenol and zearalenone levels were significantly higher in home-grown maize from areas with low rates of oesophageal cancer than from those with high rates.

In contrast, the occurrence of *F. verticillioides* in maize kernels was significantly correlated to oesophageal cancer rates. However, the rules for selection of families was different between the two populations. The prevalence of *F. verticillioides* was greater in home-grown maize collected in 12 households in the high-incidence area compared to a similar collection in a low-incidence area. Households in the high-incidence area were selected on the basis of cytological examination of cells collected from the oesophagus (Marasas et al., 1988b). Subsequent studies conducted after the chemical characterization of the fumonisins in 1988 also found significantly higher levels of *F. verticillioides* and fumonisins (20 times higher) in areas with high rates of oesophageal cancer in Transkei than in areas with low rates (Sydenham et al., 1990a,b; Rheeder et al., 1992).

F. verticillioides and *F. graminearum* cause maize ear disease under quite different ecological conditions. The environmental conditions that prevail in the areas in the Transkei with high rates of oesophageal cancer clearly favour colonization of maize ears by *F. verticillioides.* Significant fumonisin accumulation in maize occurs periodically in all such environments examined so far, primarily in relation to drought and other environmental stressors. Taking this into account, the South African studies have shown that the level of fumonisin in home-grown maize has been consistently high in the areas in the Transkei with high rates of oesophageal cancer. Cancer registry data have shown these areas to have consistently high rates of oesophageal cancer since 1955 (Jaskiewicz et al., 1987c; Makaula et al., 1996).

8.2 China

The only studies available were correlation studies where there was no clear picture on the association of either fumonisin or *F. verticillioides* contamination with oesophageal cancer.

Maize is consumed as a staple in a number of areas in China, including Linxian and Cixian counties in Henan province (Zhen, 1984; Chu & Li, 1994; Yoshizawa et al., 1994). Mortality rates for males in the high-risk areas ranged from 26 to 36 per 100 000 in the low-risk counties and from 76 to 161 per 100 000 in the high-risk counties. The incidence of *F. verticillioides* has been reported to be higher in maize in high- than low-risk areas, but the mycological data are fragmentary (Zhen, 1984) and difficult to evaluate.

Maize samples from Linxian and Cixian Counties, both high-incidence areas of oesophageal cancer in China, were analysed for FB_1 by Chu & Li (1994). All 31 samples contained FB_1 at levels ranging from 18 to 155 mg/kg. These results established that home-grown maize in high incidence areas of oesophageal cancer in China may be contaminated with very high levels of FB_1. Another investigation carried out on 246 maize samples showed that people residing in an area with high incidence of human oesophageal cancer are more exposed to fumonisins, although the exposure varied greatly (Zhang et al., 1997). However, no relationship between fumonisin and human oesophageal cancer incidence was evident from this study. In a comparative study of FB_1 levels in maize from high (Linxian) and low (Shangqiu) oesophageal cancer areas, Yoshizawa et al. (1994) found no significant differences between the areas, which was further confirmed in a subsequent study (Gao & Yoshizawa, 1997). Levels of FB_1 in 13/27 samples from the high incidence area were 0.18–2.9 (mean 0.87) mg/kg, and in 5/20 samples from the low incidence areas, FB_1 levels were 0.19–1.7 (mean 0.89) mg/kg (Yoshizawa et al., 1994).

In a comparative study of maize samples from high-risk (Haimen) and low-risk (Penlai) areas for human primary liver cancer in China, Ueno et al. (1997) reported significantly higher levels ($P < 0.01$) of total fumonisins in the high- than the low-risk area. Fumonisin levels in 80/120 samples from the high-risk area for liver cancer were 0.14–34.9 mg/kg and from the low-risk area were 0.08–15.1 mg/kg in 54/120 samples for 2 of the 3 years under investigation (Ueno et al., 1997).

8.3 Northern Italy

One analytical study was reported from Northern Italy.

Pordenone Province in the northeast of Italy has the highest mortality rate for oral and pharyngeal cancers and oesophageal cancer in Italy and amongst the highest in Europe (Franceschi et al., 1990). Risk factors identified included alcohol and tobacco use, and significant associations with maize consumption were found for oral cancer (179 cases; odds ratios 3.3; confidence intervals 2.0–5.3), pharyngeal cancer (170; 3.2; 2.0–5.3) and oesophageal cancer (68; 2.8; 1.5–5.1). There were 505 hospital controls. The elevated risk of upper digestive tract cancer was, however, limited to persons consuming more than 42 weekly drinks of alcohol (Franceschi et al., 1990). The possibility of reporting bias can not be excluded and no measures of fumonisin or *F. verticillioides* contamination were available. The analysis was restricted to men.

In this region, most maize is locally produced and eaten as cooked maize meal (polenta). Fumonisin-producing *Fusarium* species were found on maize produced in Northern Italy (Logrieco et al., 1995). One study showed that 20 samples of polenta produced in Italy in 1993 and 1994 contained 0.15–3.76 mg FB_1/kg (Pascale et al., 1995).

9. EFFECTS ON OTHER ORGANISMS IN THE LABORATORY

9.1 Microorganisms

In the only available study on the effects of fumonisins on bacteria, Becker et al. (1997) reported that FB_1 at concentrations from 50 to 1000 µM (36–721 mg/litre) did not inhibit the growth of various Gram-positive and Gram-negative bacteria. There was also no indication that FB_1 was metabolized by any of the bacteria tested.

Fumonisin was reported not to affect ethanol production (presumably by *Saccharomyces*) in distillers wash made from maize contaminated with 15 and 36 mg FB_1/kg (Bothast et al., 1992). Fumonisin concentrations of 25–100 mg/litre resulted in altered sphingolipid precursors in *Pichia ciferri* (Kaneshiro et al., 1992). This species accumulated some trihydro fatty acids in the presence of 50 mg FB_1/litre. However, in *Rhodotorula* species, FB_1 depressed production of the same compounds (Kaneshiro et al., 1993). Pure FB_1 inhibits cell growth of *Saccharomyces cerevisiae* and causes accumulation of free sphingoid bases and disruption of lipid metabolism (Wu et al., 1995).

9.2 Plants

9.2.1 Duckweed and jimsonweed

Because of their structural similarity to AAL toxins from *Alternaria alternata* f.sp. *lycopersici* (also called TA toxins; Bottini et al., 1981; Mirocha et al., 1992), fumonisins were suspected of being phytotoxic and virulence factors by several investigators. Fumonisin reduced chlorophyll synthesis by 59% in duckweed (*Lemma minor*) fronds at 10^{-6} M (Vesonder et al., 1992). Photobleaching occurred in excised jimsonweed (*Datura stramonium*) leaves, also in the µM range, and at approximately 10^{-4} M damage to mesophyll cells occurred after 6 h (Abbas et al., 1992). Fumonisin apparently causes membrane damage, as shown by electrolyte loss in jimsonweed (Abbas et al., 1991, 1993). Additionally, fumonisin disrupts the synthesis of sphingolipids in these plants (Abbas et al., 1994).

9.2.2 Tomato

FB$_1$ has similar toxicity to AAL-toxin-susceptible tomato cultivars and is not active against AAL-resistant lines. Leaf necrosis was reported at 0.4 μM by Mirocha et al. (1992) and at > 0.1 μM by Lamprecht et al. (1994). Fumonisins were reported to cause a dose-dependent reduction in shoot and root length and dry mass in tomato seedlings (Lamprecht et al., 1994). As with duckweed, fumonisin has been shown to disrupt sphingolipid metabolism in tomato (Abbas et al., 1994). In contrast, FB$_1$ added directly to excised shoots has been reported to induce callus and roots at what appear to be high doses (Bacon et al., 1994).

9.2.3 Maize

Despite reports to the contrary (Abbas & Boyette, 1992), FB$_1$ is toxic to maize cells. FB$_1$ exposure did not reduce maize seed germination but reduced radicle elongation when the solution concentration was above 10^{-4} M, and seed amylase production was inhibited (Doehlert et al., 1994). Fumonisin at concentrations in the 10 μM range decreased shoot length, shoot dry mass and root length (Lamprecht et al., 1994). FB$_1$ incorporated into plant tissue culture media reduced the growth of maize callus at 10^{-6} M (Van Asch et al., 1992). FB$_1$ has been shown to disrupt sphingolipid metabolism in maize seedlings (Riley et al., 1996). In crosses of high- and low-fumonisin-producing strains of *F. verticillioides*, only progeny that produced high concentrations of fumonisin *in vitro* caused significant stem rot (Nelson et al., 1993). These data provide indirect evidence that fumonisins play a role in the pathogenicity of *F. verticillioides* to maize (Miller, 1995).

10. FURTHER RESEARCH

- There is urgent need for an internationally available standard of pure fumonisin B_1.

- An understanding of the fate of fumonisins in maize food processing and cooking, particularly in developing countries, is urgently required.

- There is urgent need to develop a validated biomarker for human exposure to fumonisin.

- Epidemiological studies on the effects of fumonisins on human health need to be conducted, based on sound intake estimates and biomarkers.

- Valid methods for sampling for fumonisins in maize and for sampling, extracting and quantifying fumonisins in foods need to be developed.

- The influence of fumonisin on the carcinogenicity of other agriculturally important mycotoxins (e.g., aflatoxin) and carcinogenic infectious agents requires further study.

- The importance of other routes of exposure to fumonisins, including occupational exposure through inhalation, needs to be determined.

- There is urgent need for increased research on non-carcinogenic end-points including hepatotoxicity, nephrotoxicity, neurotoxicity, immunotoxicity, gastrointestinal toxicity, cardiovascular toxicity, and the mechanistic basis for the organ-selective toxicities.

- Research is needed to assess further the genotoxicity of fumonisin in both germ and somatic cells *in vitro* and *in vivo*.

- The basis for the sex differences in animals in the response to fumonisin requires further investigation.

- The environmental fate of fumonisin in the ecosystem needs to be established.

11. PREVIOUS EVALUATIONS BY INTERNATIONAL ORGANIZATIONS

The International Agency for Research on Cancer evaluated FB_1 in 1992 (IARC, 1993) and reached the following conclusions.

There is *inadequate evidence* in humans for the carcinogenicity of toxins derived from *F. verticillioides*.

There is *sufficient evidence* in experimental animals for the carcinogenicity of cultures of *F. verticillioides* that contain significant amounts of fumonisins.

There is *limited evidence* in experimental animals for the carcinogenicity of FB_1.

Overall Evaluation: Toxins derived from *Fusarium verticillioides* are *possibly carcinogenic to humans* (Group 2B).

REFERENCES

Abado-Becognee K, Mobio TA, Ennamany R, Fleurat-Lessard F, Shier WT, Badria F, & Creppy EE (1998) Cytotoxicity of fumonisin B_1: implication of lipid peroxidation and inhibition of protein and DNA syntheses. Arch Toxicol, **72**: 233-236.

Abbas HK & Boyette CD (1992) Phytotoxicity of fumonisin B_1 on weed and crop species. Weed Technol, **6**: 548-552.

Abbas HK & Ocamb CM (1995) First report of production of fumonisin B_1 by *Fusarium polyphialidicum* collected from seeds of *Pinus strobus*. Plant Dis, **79**: 642.

Abbas HK & Riley RT (1996) The presence and phytotoxicity of fumonisins and AAL-toxin in *Alternaria alternata*. Toxicon, **34**: 133-136.

Abbas HK & Shier WT (1997) Phytotoxicity of australifungin and fumonisins to weeds. In: Proceedings of the 1997 Brighton Crop Protection Conference–Weeds. Croydon, United Kingdom, British Crop Protection Council, pp 795-800.

Abbas HK, Boyette CD, Hoagland RE, & Vesonder RF (1991) Bioherbicidal potential of *Fusarium moniliforme* and its phytotoxin, fumonisin. Weed Sci, **39**: 673-677.

Abbas HK, Paul RN, Boyette CD, Duke SO, & Vesonder RF (1992) Physiological and ultrastructural effects of fumonisin on jimsonweed leaves. Can J Bot, **70**: 1824-1833.

Abbas HK, Gelderblom WCA, Cawood ME, & Shier WT (1993) Biological activities of fumonisins, mycotoxins from *Fusarium moniliforme* in jimsonweed (*Datura stramonium* L.) and mammalian cell cultures. Toxicon, **31**: 345-353.

Abbas HK, Tanaka T, Duke SO, Porter JK, Wray EM, Hodges L, Sessions AE, Wang E, Merrill AH, & Riley RT (1994) Fumonisin- and AAL-toxin-induced disruption of sphingolipid metabolism with accumulation of free sphingoid bases. Plant Physiol, **106**: 1085-1093.

Abbas HK, Ocamb CM, Xie W, Mirocha CJ, & Shier WT (1995) First report of fumonisin B_1, B_2, and B_3 production by *Fusarium oxysporum* var. *redolens*. Plant Dis, **79**: 968.

Abbas HK, Cartwright RD, Shier WT, Abouzied MM, Bird CB, Rice LG, Ross PF, Sciumbato GL, & Meredith FI (1998) Natural occurrence of fumonisins in rice with *Fusarium* sheat rot disease. Plant Disease, **82**: 22-25.

Abel S & Gelderblom WCA (1998) Oxidative damage and fumonisin B_1-induced toxicity in primary rat hepatocytes and rat liver *in vivo*. Toxicology, **131**(2-3): 121-131.

Alberts JF, Gelderblom WCA, Thiel PG, Marasas WFO, Van Schalkwyk DJ, & Behrend Y (1990) Effects of temperature and incubation period on production of fumonisin B_1 by *Fusarium moniliforme*. Appl Environ Microbiol, **56**: 1729-1733.

Alberts JF, Gelderblom WCA, Vleggaar R, Marasas WFO, & Rheeder JP (1993) Production of ^{14}C-labelled fumonisin B_1 by *Fusarium moniliforme* MRC 826 in corn culture. Appl Environ Microbiol, **59**: 2673-2677.

Ali NS, Yamashita A, & Yoshizawa T (1998) Natural co-occurrence of aflatoxins and *Fusarium* mycotoxins (fumonisins, deoxynivalenol, nivalenol and zearalenone) in corn from Indonesia. Food Addit Contam, **15**: 377-384.

90

Arora A, Jones BJ, Patel TC, Bronk SF, & Gores GJ (1997) Ceramide induces hepatocyte cell death through disruption of mitochondrial function in the rat. Hepatology, **25**: 958-963.

Azcona-Olivera JI, Abouzied MM, Plattner RD, Norred WP, & Pestka JJ (1992a) Generation of antibodies reactive with fumonisins B_1, B_2 and B_3 by using cholera toxin as the carrier-adjuvant. Appl Environ Microbiol, **58**: 169-173.

Azcona-Olivera JI, Abouzied MM, Plattner RD, & Pestka JJ (1992b) Production of monoclonal antibodies to the mycotoxins fumonisin B_1, B_2, and B_3. J Agric Food Chem, **40**: 531-534.

Azuka CE, Osweiler GD, & Reynolds DL (1993) Effect of glutathione and vitamin E on cytotoxicity of fumonisins. Toxicologist, **13**: 233.

Bacon CW & Williamson JW (1992) Interactions of *Fusarium moniliforme*, its metabolites and bacteria with corn. Mycopathologia, **117**: 65-71.

Bacon CW, Hinton DM, Chamberlain WJ, & Norred WP (1994) *De novo* induction of adventitious roots in excised shoots of tomatoes by fumonisin B_1, a metabolite of *Fusarium moniliforme*. J Plant Growth Regul, **13**: 53-57.

Bacon CW, Porter JK, & Norred WP (1995) Toxic interaction of fumonisin B_1 and fusaric acid measured by injection into fertile chicken egg. Mycopathologia, **129**: 29-35.

Badiani K, Byers DM, Cook HW, & Ridgway ND (1996) Effect of fumonisin B_1 on phosphatidylethanolamine biosynthesis in Chinese hamster ovary cells. Biochim Biophys Acta, **1304**: 190-196.

Ballou LR, Laulederkind SJF, Rosloniec EF, & Raghow R (1996) Ceramide signalling and the immune response. Biochim Biophys Acta, **1301**: 273-287.

Balsinde J, Balboa MA, & Dennis EA (1997) Inflammatory activation of arachidonic acid signaling in murine P388D1 macrophages via sphingomyelin synthesis. J Biol Chem, **272**: 20373-20377.

Bane DP, Neumann EJ, Hall WF, Harlin KS, & Slife LN (1992) Relationship between fumonisin contamination of feed and mystery swine disease. Mycopathologia, **117**: 121-124.

Becker BA, Pace L, Rottinghaus GE, Shelby R, Misfeldt M, & Ross MS (1995) Effects of feeding fumonisin B_1 in lactating sows and their suckling pigs. Am J Vet Res, **56**: 1253-1258.

Becker B, Bresch H, Schillinger U, & Thiel PG (1997) The effect of fumonisin B_1 on the growth of bacteria. World J Microbiol Biotechnol, **13**: 539-543.

Bell RM, Hannun YA, & Merrill AH Jr ed. (1993) Advances in lipid research: sphingolipids and their metabolites. Orlando, Florida, Academic Press, vol 25-26.

Bennett GA & Richard JL (1994) Liquid chromatographic method for the naphthalene dicarboxaldehyde derivate of fumonisins. J Assoc Off Anal Chem Int, **77**: 501-505.

Bennett GA & Richard JL (1996) Influence of processing on *Fusarium* mycotoxins in contaminated grains. Food Technol, **May**: 235-238.

Bermudez AJ, Ledoux DR, Turk JR, & Rottinghaus GE (1996) The chronic effects of *Fusarium moniliforme* culture material containing known levels of fumonisin B₁, in turkeys. Avian Dis, **40**: 231-235.

Bermudez AJ, Ledoux DR, Rottinghaus GE, & Bennett GA (1997) The individual and combined effects of the *Fusarium* mycotoxins moniliformin and fumonisin B₁ in turkeys. Avian Dis, **41**: 304-311.

Bezuidenhout SC, Gelderblom WCA, Gorst-Allman CP, Horak RM, Marasas WFO, Spiteller G, & Vleggaar R (1988) Structure elucidation of the fumonisins, mycotoxins from *Fusarium moniliforme*. J Chem Soc Chem Commun, **1988**: 743-745.

Bhat RV, Shetty PH, Amruth RP, & Sudershan RV (1997) A foodborne disease outbreak due to the consumption of moldy sorghum and maize containing fumonisin mycotoxins. Clin Toxicol, **35**: 249-255.

Blackwell BA, Miller JD, & Savard ME (1994) Production of carbon 14-labeled fumonisin in liquid culture. J Assoc Off Anal Chem Int, **77**: 506-511.

Boland MP, Foster SJ, & O'Neill LAJ (1997) Daunorubicin activates NFkB and induces kB-dependent gene expression in HL-60 promyelocytic and Jurkat T lymphoma cells. J Biol Chem, **272**: 12952-12960.

Boldin S & Futerman AH (1997) Glucosylceramide synthesis is required for basic fibroblast growth factor and laminin to stimulate axonal growth. J Neurochem, **68**: 882-885.

Bondy G, Suzuki C, Barker M, Armstrong C, Fernie S, Hierlihy L, Rowsell P, & Mueller R (1995) Toxicity of fumonisin B₁ administered intraperitoneally to male Sprague-Dawley rats. Food Chem Toxicol, **33**: 653-665.

Bondy GS, Suzuki CAM, Fernie SM, Armstrong CL, Hierlihy SL, Savard ME, & Barker MG (1997) Toxicity of fumonisin B₁ to B6C3F₁ mice: a 14-day gavage study. Food Chem Toxicol, **35**: 981-989.

Bondy GS, Suzuki CAM, Mueller RW, Fernie SM, Armstrong CL, Hierlihy SL, Savard ME, & Barker MG (1998) Gavage administration of the fungal toxin fumonisin B₁ to female Sprague-Dawley rats. J Toxicol Environ Health, **53**: 135-151.

Bose R, Verheij M, Haimovitz-Friedman A, Scotto K, Fuks Z, & Kolesnick R (1995) Ceramide synthase mediates daunorubicin-induced apoptosis: An alternative mechanism for generating death signals. Cell, **82**: 404-414.

Bothast RJ, Bennett GA, Vancauwenberge JE, & Richard JL (1992) Fate of fumonisin B₁ in naturally contaminated corn during ethanol fermentation. Appl Environ Microbiol, **58**: 233-236.

Bottalico A, Logrieco A, Ritieni A, Moretti A, Randazzo G, & Corda P (1995) Beauvericin and fumonisin B₁ in preharvest *Fusarium moniliforme* maize ear rot in Sardinia. Food Addit Contam, **12**: 599-607.

Bottini AT, Bowen JR, & Gilchrist DG (1981) Phytotoxins: II. Characterization of a phytotoxic fraction from *Alternaria alternata* f. sp. *lycopersici*. Tetrahedron Lett, **22**: 2723-2726.

Bradlaw J, Pritchard D, Flynn T, Eppley R, & Stack M (1994) *In vitro* assessment of fumonisin B₁ toxicity using reaggregate cultures of chick embryo neural retina cells (CERC). In Vitro Cell Dev Biol, **30A**: 92.

Brown DW, McCoy CP, & Rottinghaus GE (1994) Experimental feeding of *Fusarium moniliforme* culture material containing fumonisin B₁ to channel catfish, *Ictalurus punctatus*. J Vet Diagn Invest, **6**: 123-124.

Brown TP, Rottinghaus GE, & Williams ME (1992) Fumonisin mycotoxicosis in broilers: Performance and pathology. Avian Dis, **36**: 450-454.

Bryden WL, Love RJ, & Burgess LW (1987) Feeding grain contaminated with *Fusarium graminearum* and *Fusarium moniliforme* to pigs and chickens. Aust Vet J, **64**: 225-226.

Bryden WL, Ravindran G, Amba MT, Gill RJ, & Burgess LW (1996) Mycotoxin contamination of maize grown in Australia, the Philippines and Vietnam. In: Miraglia M, Brera C, & Onori R ed. Ninth International IUPAC Symposium on Mycotoxins and Phycotoxins, Rome, 27-31 May 1996: Abstract book. Rome, Istituto Superiore di Sanità, p 41.

Bucci TJ, Hansen DK, & LaBorde JB (1996) Leukoencephalomalacia and hemorrhage in the brain of rabbits gavaged with mycotoxin fumonisin B₁. Nat Toxins, **4**: 51-52.

Bullerman LB & Tsai W-Y (1994) Incidence and levels of *Fusarium moniliforme, Fusarium proliferatum* and fumonisins in corn and corn-based foods and feeds. J Food Prot, **57**: 541-546.

Burdaspal PA & Legarda TM (1996) Occurrence of fumonisins in corn and processed corn-based commodities for human consumption in Spain. In: Miraglia M, Brera C, & Onori R ed. IX International IUPAC Symposium on Mycotoxins and Phycotoxins, Rome, 27-31 May 1996: Abstract book. Rome, Istituto Superiore di Sanità, p 167.

Butler T (1902) Notes on a feeding experiment to produce leucoencephalitis in a horse, with positive results. Am Vet Rev, **26**: 748-751.

Cabot MC, Han T-Y, & Giuliano AE (1998) The multidrug resistance modulator SDZ PSC 833 is a potent activator of cellular ceramide formation. FEBS Lett, **431**: 185-188.

Cabot MC, Giuliano AE, Han T-Y, & Liu Y-Y (1999) SDZ PSC 833, the cyclosporine A analogue and multidrug resistance modulator, activates ceramide synthesis and increases vinblastine sensitivity in drug-sensitive and drug-resistant cancer cells. Cancer Res, **59**: 880-885.

Caramelli M, Dondo A, Cantini Cortellazzi G, Visconti A, Minervini F, Doko MB, & Guarda F (1993) [Equine leukoencephalomalacia from fumonisin: First case in Italy.] Ippologia, **4**: 49-56 (in Italian).

Casteel SW, Turk JR, Cowart RP, & Rottinghaus GE (1993) Chronic toxicity of fumonisin in weanling pigs. J Vet Diagn Invest, **5**: 413-417.

Casteel SW, Turk JR, & Rottinghaus GE (1994) Chronic effects of dietary fumonisin on the heart and pulmonary vasculature of swine. Fundam Appl Toxicol, **23**: 518-524.

Castegnaro M, Garren L, Gaucher I, & Wild CP (1996) Development of a new method for the analysis of sphinganine and sphingosine in urine and tissues. Nat Toxins, **4**: 284-290.

Castella G, Bragulat MR, & Cabañes FJ (1997) Occurrence of *Fusarium* species and fumonisins in some animal feeds and raw materials. Cereal Res Commun, **25**: 355-356.

Cawood ME, Gelderblom WCA, Vleggaar R, Behrend Y, Thiel PG, & Marasas WFO (1991) Isolation of the fumonisin mycotoxins: A quantitative approach. J Agric Food Chem, **39**: 1958-1962.

Cawood ME, Gelderblom WCA, Alberts JF, & Snyman SD (1994) Interaction of ^{14}C-labelled fumonisin B mycotoxins with primary rat hepatocyte cultures. Food Chem Toxicol, **32**: 627-632.

Chamberlain WJ, Voss KA, & Norred WP (1993) Analysis of commercial laboratory rat rations for fumonisin B₁, a mycotoxin produced on corn by *Fusarium moniliforme*. Contemp Top, **32**: 26-28.

Chelkowski J & Lew H (1992) *Fusarium* species of Liseola section - occurrence in cereals and ability to produce fumonisins. Microbiol Alim Nutr, **10**: 49-53.

Chu FS & Li GY (1994) Simultaneous occurrence of fumonisin B₁ and other mycotoxins in moldy corn collected from the People's Republic of China in regions with high incidences of esophageal cancer. Appl Environ Microbiol, **60**: 847-852.

Chulze SN, Ramirez ML, Farnochi MC, Pascale M, Visconti A, & March G (1996) *Fusarium* and fumonisin occurrence in Argentinian corn at different ear maturity stages. J Agric Food Chem, **44**: 2797-2801.

Churchwell MI, Cooper WM, Howard PC, & Doerge DR (1997) Determination of fumonisins in rodent feed using HPLC with electrospray mass spectrometric detection. J Agric Food Chem, **45**: 2573-2578.

Ciacci-Zanella JR & Jones C (1999) Analysis of apoptosis induced by fumonisin B₁, a novel mycotoxin frequently found in cereal grains. Food Chem Toxicol, **37**: 703-712.

Ciacci-Zanella JR, Merrill AH Jr, Wang E, & Jones C (1998) Characterization of cell cycle arrest by fumonisin B₁ in CV-1 cells. Food Chem Toxicol, **36**: 791-804.

Collins TF, Shacklelford ME, Sprando RL, Black TN, LaBorde JB, Hansen DK, Eppley RM, Trucksess MW, Howard PC, Bryant MA, Ruggles DI, Olejnik N, & Rorie JI (1998a) Effects of fumonisin B₁ in pregnant rats. Food Chem Toxicol, **36**: 397-408.

Collins TF, Sprando RL, Black TN, Shacklelford ME, LaBorde JB, Hansen DK, Eppley RM, Trucksess MW, Howard PC, Bryant MA, Ruggles DI, Olejnik N, & Rorie JI (1998b) Effects of fumonisins in pregnant rats, part 2. Food Chem Toxicol, **36**: 673-685.

Colvin BM & Harrison LR (1992) Fumonisin-induced pulmonary edema and hydrothorax in swine. Mycopathologia, **117**: 79-82.

Colvin BM, Cooley AJ, & Beaver RW (1993) Fumonisin toxicosis in swine: Clinical and pathologic findings. J Vet Diagn Invest, **5**: 232-241.

Counts RS, Nowak G, Wyatt RD, & Schnellmann RG (1996) Nephrotoxicant inhibition of renal proximal tubule cell regeneration. Am J Physiol, **269**: F274-F281.

Dawlatana M, Coker RD, Nagler MJ, & Blunden G (1995) A normal phase HPTLC method for the quantitative determination of fumonisin B_1 in rice. Chromatographia, **41**: 187-190.

De León C & Pandey S (1989) Improvement of resistance to ear and stalk rots and agronomic traits in tropical maize gene pools. Crop Sci, **29**: 12-17.

de Nijs M (1998) Public health aspects of *Fusarium* mycotoxins in food in The Netherlands - a risk assessment. Wageningen, The Netherlands, Agricultural University, 140 pp (Thesis).

de Nijs M, van Egmond HP, Nauta M, Rombouts FM, & Notermans SHW (1998a) Assessment of human exposure to fumonisin B_1. J Food Prot, **61**: 879-884.

de Nijs M, Sizoo EA, Rombouts FM, Notermans SHW, & van Egmond HP (1998b) Fumonisin B_1 in maize for food production imported in The Netherlands. Food Addit Contam, **15**: 389-392.

de Nijs M, Sizoo EA, Vermunt AEM, Notermans SHW, & van Egmond HP (1998c) The occurrence of fumonisin B_1 in maize-containing foods in The Netherlands. Food Addit Contam, **15**: 385-388.

Desjardins AE, Plattner RD, & Nelson PE (1994) Fumonisin production and other traits of *Fusarium moniliforme* strains from maize in northeast Mexico. Appl Environ Microbiol, **60**: 1695-1697.

Desjardins AE, Plattner RD, Lu M, & Claflin LE (1998) Distribution of fumonisins in maize ears infected with strains of *Fusarium moniliforme* that differ in fumonisin production. Plant Dis, **82**: 953-958.

Diaz GJ & Boermans HJ (1994) Fumonisin toxicosis in domestic animals: a review. Vet Hum Toxicol, **36**: 548-555.

DiPietrantonio AM, Hsieh T, Olson SC, & Wu JM (1998) Regulation of G1/S transition and induction of apoptosis in HL-60 leukemia cells by fenretinide (4HPR). Int J Cancer, **78**: 53-61.

Doehlert DC, Knutson CA, & Vesonder RF (1994) Phytotoxic effects of fumonisin B_1 on maize seedling growth. Mycopathologia, **127**: 117-121.

Doerge DR, Howard PC, Bajic S, & Preece S (1994) Determination of fumonisins using on-line liquid chromatography coupled to electrospray mass spectrometry. Rapid Commun Mass Spectrom, **8**: 603-606.

Doko MB & Visconti A (1994) Occurrence of fumonisins B_1 and B_2 in corn and corn-based human foodstuffs in Italy. Food Addit Contam, **11**: 433-439.

Doko MB, Rapior S, & Visconti A (1994) Screening for fumonisins B_1 and B_2 in corn and corn-based foods and feeds from France. In: Abstracts of the 7th International Congress of the IUMS Mycology Division, Prague, Czech Republic, 3-8 July 1994, p 468.

Doko MB, Rapior S, Visconti A, & Schjøth JE (1995) Incidence and levels of fumonisin contamination in maize genotypes grown in Europe and Africa. J Agric Food Chem, **43**: 429-434.

Doko MB, Canet C, Brown N, Sydenham EW, Mpuchane S, & Siame BA (1996) Natural co-occurrence of fumonisins and zearalenone in cereals and cereal-based foods from Eastern and Southern Africa. J Agric Food Chem, **44**: 3240-3243.

Dombrink-Kurtzman MA, Javid T, Bennett GA, Richard JL, Cote LM, & Buck WB (1993) Lymphocyte cytotoxicity and erythrocytic abnormalities induced in broiler chicks by fumonisins B₁ and B₂ and moniliformin from *Fusarium proliferatum*. Mycopathologia, **124**: 47-54.

Dombrink-Kurtzman MA, Bennett GA, & Richard JL (1994a) An optimized MTT bioassay for determination of cytotoxicity of fumonisins in turkey lymphocytes. J Assoc Off Anal Chem Int, **77**: 512-516.

Dombrink-Kurtzman MA, Bennett GA, & Richard JL (1994b) Induction of apoptosis in turkey lymphocytes by fumonisin. FASEB J, **8**: A488.

Dragoni I, Pascale M, Piantanida L, Tirilly Y, & Visconti A (1996) [Presence of fumonisin in foodstuff destined for feeding to pigs in Britanny (France).] Microbiol Alim Nutr, **14**: 97-103 (in Italian).

Dugyala RR, Sharma RP, Tsunoda M, & Riley RT (1998) Tumor necrosis factor-*alpha* as a contributor in fumonisin B₁ toxicity. J Pharmacol Exp Ther, **285**: 317-324.

Duncan K, Kruger S, Zabe N, Kohn B, & Prioli R (1998) Improved fluorometric and chromatographic methods for the quantification of fumonisins B₁, B₂ and B₃. J Chromatogr, **A815**: 41-47.

Dupuy J, Le Bars P, Boudra H, & Le Bars J (1993a) Thermostability of fumonisin B₁, a mycotoxin from *Fusarium moniliforme*, in corn. Appl Environ Microbiol, **59**: 2864-2867.

Dupuy J, Le Bars P, Le Bars J, & Boudra H (1993b) Determination of fumonisin B₁ in corn by instrumental thin layer chromatography. J Planar Chromatogr, **6**: 476-480.

Dutton MF (1996) Fumonisins, mycotoxins of increasing importance: their nature and their effects. Pharmacol Ther, **70**: 137-161.

Duvick J, Rood T, & Grant S (1994) Isolation of fumonisin-metabolizing fungi from maize seed. In: Proceedings of the Fifth International Mycological Congress, Vancouver, Canada, 14-21 August 1994, 56 pp.

Duvick J, Rood T, Maddox J, & Gilliam J (1998) Detoxification of mycotoxins in plants as a strategy for improving grain quality and disease resistance: Identification of fumonisin-degrading microbes from maize. In: Kohmoto K & Yoder OC ed. Molecular genetics of host specific toxins in plant diseases. Dordrecht, Kluwer Academic Publishers, pp 369-381.

Edrington TS, Kamps-Holtzapple CA, Harvey RB, Kubena LF, Elissalde MH, & Rottinghaus GE (1995) Acute hepatic and renal toxicity in lambs dosed with fumonisin-containing culture material. J Anim Sci, **73**: 508-515.

Espada Y, Ruiz de Gopegui R, Cuadradas C, & Cabañes FJ (1994) Fumonisin mycotoxicosis in broilers: Weights and serum chemistry modifications. Avian Dis, **38**: 454-460.

Espada Y, Ruiz de Gopegui R, Cuadradas C, & Cabañes FJ (1997) Fumonisin mycotoxicosis in broilers: Plasma proteins and coagulation modifications. Avian Dis, **41**: 73-79.

Farrar JJ & Davis RM (1991) Relationships among ear morphology, western flower thrips and *Fusarium* ear rot of corn. Phytopathology, **81**: 661-666.

Fazekas B, Bajmócy E, Glávits R, Fenyvesi A, & Tanyi J (1998) Fumonisin B_1 contamination of maize and experimental acute fumonisin toxicosis in pigs. J Vet Med, **B45**: 171-181.

Ferguson SA, Omer VE, Kwon OS, Holson RR, Houston RJ, Rottinghaus GE, & Slikker W Jr (1997) Prenatal fumonisin (FB_1) treatment in rats results in minimal maternal or offspring toxicity. Neurotoxicology, **18**: 561-569.

Fincham JE, Marasas WFO, Taljaard JJF, Kriek NPJ, Badenhorst CJ, Gelderblom WCA, Seier JV, Smuts CM, Faber M, Weight MJ, Slazus W, Woodroof CW, van Wyk MJ, Kruger M, & Thiel PG (1992) Atherogenic effects in a non-human primate of *Fusarium moniliforme* cultures added to a carbohydrate diet. Atherosclerosis, **94**: 13-25.

Floss JL, Casteel SW, Johnson GC, & Rottinghaus GE (1994a) Developmental toxicity in hamsters of an aqueous extract of *Fusarium moniliforme* culture material containing known quantities of fumonisin B_1. Vet Hum Toxicol, **36**: 5-10.

Floss JL, Casteel SW, Johnson GC, Rottinghaus GE, & Krause GF (1994b) Developmental toxicity of fumonisin in Syrian hamsters. Mycopathologia, **128**: 33-38.

Flynn GL (1985) Mechanism of percutaneous absorption from physicochemical evidence. In: Bronaugh RL & Maibach HI ed. Percutaneous absorption. New York, Marcel Dekker, pp 17-42.

Flynn TJ, Pritchard D, Bradlaw J, Eppley R, & Page S (1994) Effects of the mycotoxin fumonisin B_1 and its alkaline hydrolysis product on pre-somite rat embryos *in vitro*. Teratology, **49**: 404.

Flynn TJ, Pritchard D, Bradlaw JA, Eppley R, & Page S (1996) *In vitro* embryotoxicity of fumonisin B_1 evaluated with cultured postimplantation staged rat embryos. Toxicol In Vitro, **9**: 271-279.

Flynn TJ, Stack ME, Troy AL, & Chirtel SJ (1997) Assessment of the embryotoxic potential of the total hydrolysis product of fumonisin B_1 using cultured organogenesis-staged rat embryos. Food Chem Toxicol, **35**: 1135-1141.

Franceschi S, Bidoli E, Barón AE, & La Vecchia C (1990) Maize and risk of cancer of the oral cavity, pharynx and esophagus in northeastern Italy. J Natl Cancer Inst, **82**: 1407-1411.

Fukuda H, Shima H, Vesonder RF, Tokuda H, Nishino H, Katoh S, Tamura S, Sugimura T, & Nagao M (1996) Inhibition of protein serine/threonine phosphatases by fumonisin B_1, a mycotoxin. Biochem Biophys Res Commun, **220**: 160-165.

97

Furuya S, Ono K, & Hirabayashi Y (1995) Sphingolipid biosynthesis is necessary for dendrite growth and survival of cerebellar Purkinje cells in culture. J Neurochem, **65**: 1551-1561.

Gao H-P &Yoshizawa T (1997) Further study on *Fusarium* mycotoxins in corn and wheat from a high-risk area for human esophageal cancer in China. Mycotoxins, **45**: 51-55.

Garzotto M, White-Jones M, Jiang Y, Ehleiter D, Liao W-C, Haimovitz-Friedman A, Fuks Z, & Kolesnick R (1998) 12-O-tetradecanoylphorbol-13-acetate-induced apoptosis in LNCaP cells is mediated through ceramide synthase. Cancer Res, **58**: 2260-2264.

Gelderblom WCA & Snyman SD (1991) Mutagenicity of potentially carcinogenic mycotoxins produced by *Fusarium moniliforme*. Mycotoxin Res, **7**: 46-52.

Gelderblom WCA, Jaskiewicz K, Marasas WFO, Thiel PG, Horak RM, Vleggaar R, & Kriek NPJ (1988) Fumonisins - Novel mycotoxins with cancer-promoting activity produced by *Fusarium moniliforme*. Appl Environ Microbiol, **54**: 1806-1811.

Gelderblom W, Marasas W, Thiel P, Semple E, & Farber E (1989) Possible non-genotoxic nature of active carcinogenic components produced by *Fusarium moniliforme*. Proc Am Assoc Cancer Res, **30**: 144.

Gelderblom WCA, Kriek NPJ, Marasas WFO, & Thiel PG (1991) Toxicity and carcinogenicity of the *Fusarium moniliforme* metabolite, fumonisin B₁, in rats. Carcinogenesis, **12**: 1247-1251.

Gelderblom WCA, Marasas WFO, Vleggaar R, Thiel PG, & Cawood ME (1992a) Fumonisins: Isolation, chemical characterization and biological effects. Mycopathologia, **117**: 11-16.

Gelderblom WCA, Semple E, Marasas WFO, & Farber E (1992b) The cancer-initiating potential of the fumonisin B mycotoxins. Carcinogenesis, **13**: 433-437.

Gelderblom WCA, Cawood ME, Snyman SD, Vleggaar R, & Marasas WFO (1993) Structure-activity relationships of fumonisins in short-term carcinogenesis and cytotoxicity assays. Food Chem Toxicol, **31**: 407-414.

Gelderblom WCA, Cawood ME, Snyman SD, & Marasas WFO (1994) Fumonisin B₁ dosimetry in relation to cancer initiation in rat liver. Carcinogenesis, **15**: 209-214.

Gelderblom WCA, Snyman SD, van der Westhuizen L, & Marasas WFO (1995) Mitoinhibitory effect of fumonisin B₁ on rat hepatocytes in primary culture. Carcinogenesis, **16**: 625-631.

Gelderblom WCA, Smuts CM, Abel S, Snyman SD, Cawood MA, van der Westhuizen L, & Swanevelder S (1996a) Effect of fumonisin B₁ on protein and lipid synthesis in primary rat hepatocytes. Food Chem Toxicol, **34**: 361-369.

Gelderblom WCA, Snyman SD, Abel S, Lebepe-Mazur S, Smuts CM, van der Westhuizen L, Marasas WFO, Victor TC, Knasmüller S, & Huber W (1996b) Hepatotoxicity and carcinogenicity of the fumonisins in rats: A review regarding mechanistic implications for establishing risk in humans. Adv Exp Med Biol, **392**: 279-296.

Gelderblom WCA, Snyman SD, Lebepe-Mazur S, van der Westhuizen L, Kriek NPJ, & Marasas WFO (1996c) The cancer-promoting potential of fumonisin B_1 in rat liver using diethylnitrosamine as a cancer initiator. Cancer Lett, **109**: 101-108.

Gelderblom WCA, Smuts CM, Abel S, Snyman SD, van der Westhuizen L, Huber WW, & Swanevelder S (1997) Effect of fumonisin B_1 on the levels and fatty acid composition of selected lipids in rat liver, *in vivo*. Food Chem Toxicol, **35**: 647-656.

Gillard BK, Harrell RG, & Marcus DM (1996) Pathways of glycosphingolipid biosynthesis in SW13 cells in the presence and absence of vimentin intermediate filaments. Glycobiology, **6**: 33-42.

Goel S, Lenz SD, Lumlertdacha S, Lovell RT, Shelby RA, Li M, Riley RT, & Kemppainen BW (1994) Sphingolipid levels in catfish consuming *Fusarium moniliforme* corn culture material containing fumonisins. Aquat Toxicol, **30**: 285-294.

Gross SM, Reddy RV, Rottinghaus GE, Johnson G, & Reddy CS (1994) Developmental effects of fumonisin B_1-containing *Fusarium moniliforme* culture extract in CD1 mice. Mycopathologia, **128**: 111-118.

Gumprecht LA, Marcucci A, Weigel RM, Vesonder RF, Riley RT, Showker JL, Beasley VR, & Haschek WM (1995) Effects of intravenous fumonisin B_1 in rabbits: nephrotoxicity and sphingolipid alterations. Nat Toxins, **3**: 395-403.

Gumprecht LA, Beasley VR, Weigel RM, Parker HM, Tumbleson ME, Bacon CW, Meredith FI, & Haschek WM (1998) Development of fumonisin-induced hepatotoxicity and pulmonary edema in orally dosed swine: morphological and biochemical alterations. Toxicol Pathol, **26**: 777-788.

Guzman RE, Bailey K, Casteel SW, Turk J, & Rottinghaus GE (1997) Dietary *Fusarium moniliforme* culture material induces *in vitro* tumor necrosis factor-alpha like activity in sera of swine. Immunopharmacol Immuntoxicol, **19**: 279-289.

Hall JO, Javed T, Bennett GA, Richard JL, Dombrink-Kurtzman MA, Côté LM, & Buck WB (1995) Serum vitamin A (retinol) reduction in broiler chicks on feed amended with *Fusarium proliferatum* culture material or fumonisin B_1 and moniliformin. J Vet Diagn Invest, **7**: 416-418.

Hammer P, Blüthgen A, & Walte HG (1996) Carry-over of fumonisin B_1 into the milk of lactating cows. Milchwissenschaft, **51**: 691-695.

Hanada K, Nishijima M, & Akamatsu Y (1990) A temperature-sensitive mammalian cell mutant with thermolabile serine palmitoyltransferase for the sphingolipid biosynthesis. J Biol Chem, **265**: 22137-22142.

Hanada K, Nishijima M, Kiso H Jr, Hasegawa A, Fujita S, Ogawa T, & Akamatsu Y (1992) Sphingolipids are essential for the growth of Chinese hamster ovary cells: Restoration of the growth of a mutant defective in sphingoid base biosynthesis by exogenous sphingolipids. J Biol Chem, **267**: 23527-23533.

Hanada K, Izawa K, Nishijima M, & Akamatsu Y (1993) Sphingolipid deficiency induces hypersensitivity of CD14, a glycosyl phosphatidylinositol-anchored protein, to phosphatidylinositol-specific phospholipase C. J Biol Chem, **268**: 13820-13823.

Hannun YA, Merrill AH Jr, & Bell RM (1991) Use of sphingosine as an inhibitor of protein kinase C. Meth Enzymol, **201**: 316-328.

Harel R & Futerman AH (1993) Inhibition of sphingolipid synthesis affects axonal outgrowth in cultured hippocampal neurons. J Biol Chem, **268**: 14476-14481.

Harrison LR, Colvin BM, Greene JT, Newman LE, & Cole JR Jr (1990) Pulmonary edema and hydrothorax in swine produced by fumonisin B₁, a toxic metabolite of *Fusarium moniliforme*. J Vet Diagn Invest, **2**: 217-221.

Hartwell LH & Kastan MB (1994) Cell cycle control and cancer. Science, **266**: 1821-1828.

Harvey RB, Edrington TS, Kubena LF, Elissalde MH, & Rottinghaus GE (1995) Influence of aflatoxin and fumonisin B₁-containing culture material on growing barrows. Am J Vet Res, **56**: 1668-1672.

Harvey RB, Edrington TS, Kubena LF, Elissalde MH, Casper HH, Rottinghaus GE, & Turk JR (1996) Effects of dietary fumonisin B₁-containing culture material, deoxynivalenol-contaminated wheat, or their combination on growing barrows. Am J Vet Res, **57**: 1790-1794.

Haschek WM, Motelin G, Ness DK, Harlin KS, Hall WF, Vesonder R, Peterson RE, & Beasley VR (1992) Characterization of fumonisin toxicity in orally and intravenously dosed swine. Mycopathologia, **117**: 83-96.

Haschek WM, Gumprecht LA, Smith GW, Parker HM, Beasley VR, & Tumbleson ME (1996) Effects of fumonisins in swine. In: Advances in swine biomedical research. New York, Plenum Press, pp 99-112.

Hendrich S, Miller KA, Wilson TM, & Murphy PA (1993) Toxicity of *Fusarium proliferatum*-fermented nixtamalized corn-based diets fed to rats: Effect of nutritional status. J Agric Food Chem, **41**: 1649-1654.

Hendricks K (1999) Fumonisins and neural tube defects in South Texas. Epidemiology, **10**: 198-200.

Henry MH (1993) Bioproduction of fumonisin B₁ in chicken embryos and broiler chicks. University of Georgia, 93 pp (Masters of Science Thesis).

Hesseltine CW, Rogers RF, & Shotwell OL (1981) Aflatoxin and mold flora in North Carolina in 1977 corn crop. Mycologia, **73**: 216-228.

Hidari KI-PJ, Ichikawa S, Fujita T, Sakiyama H, & Hirabayashi Y (1996) Complete removal of sphingolipids from the plasma membrane disrupts cell to substratum adhesion of mouse melanoma cells. J Biol Chem, **271**: 14636-14641.

Hirooka EY, Yamaguchi MM, Aoyama S, Sugiura Y, & Ueno Y (1996) The natural occurrence of fumonisins in Brazilian corn kernels. Food Addit Contam, **13**: 173-183.

Hoenisch RW & Davis RM (1994) Relationship between kernel pericarp thickness and susceptibility to *Fusarium* ear rot in field corn. Plant Dis, **78**: 517-519.

Holcomb M & Thompson HC Jr (1996) Analysis of fumonisin B_1 in rodent feed by CE with fluorescence detection of the FMOC derivative. J Cap Electrophor, **3**: 205-208.

Holcomb M, Thompson HC Jr, & Hankins LJ (1993) Analysis of fumonisin B_1 in rodent feed by gradient elution HPLC using precolumn derivatization with FMOC and fluorescence detection. J Agric Food Chem, **41**: 764-767.

Hopmans EC & Murphy PA (1993) Detection of fumonisins B_1, B_2, and B_3 and hydrolyzed fumonisin B_1 in corn-containing foods. J Agric Food Chem, **41**: 1655-1658.

Hopmans EC, Hauck CC, Hendrich S, & Murphy PA (1997) Excretion of fumonisin B_1, hydrolyzed fumonisin B_1, and the fumonisin FB_1-fructose adduct in rats. J Agric Food Chem, **45**: 2618-2625.

Horvath A, Sutterlin C, Manning-Krieg U, Movva NR, & Riezman H (1994) Ceramide synthesis enhances transport of GPI-anchored proteins to the Golgi apparatus in yeast. EMBO J, **13**:3687-3695.

Howard PC, Churchwell MI, Couch LH, Marques MM, & Doerge DR (1998) Formation of N-(carboxymethyl)fumonisin B_1, following the reaction of fumonisin B_1 with reducing sugars. J Agric Food Chem, **46**: 3546-3557.

Howard PC, Couch LH, Muskhelishvili M, Eppley RM, Doerge DR, & Okerberg C (1999) Comparative toxicity of fumonisin derivatives in 28-day feeding study using female $B6C3F_1$ mice. Toxicologist, **48**: 34.

Huang C, Dickman M, Henderson G, & Jones C (1995) Repression of protein kinase C and stimulation of cyclic AMP response elements by fumonisin, a fungal encoded toxin which is a carcinogen. Cancer Res, **55**: 1655-1659.

Humpf H-U, Schmelz E-M, Meredith FI, Vesper H, Vales TR, Wang E, Menaldino DS, Liotta DC, & Merrill AH Jr (1998) Acylation of naturally occurring and synthetic 1-deoxysphinganines by ceramide synthase: Formation of N-palmitoyl-aminopentol produces a toxic metabolite of hydrolyzed fumonisin, AP_1, and a new category of ceramide synthase inhibitor. J Biol Chem, **273**: 19060-19064.

Humphreys SH, Carrington C, & Bolger PM (1997) Fumonisin risk scenarios. Toxicologist, **36**: 170.

IARC (1993) Toxins derived from *Fusarium moniliforme*: Fumonisins B_1 and B_2 and fusarin C. In: Some naturally occurring substances: Food items and constituents, heterocyclic aromatic amines and mycotoxins. Lyon, International Agency for Research on Cancer, pp 445-466 (IARC Monographs on the Evaluation of Carcinogenic Risk to Humans, Volume 56).

Inooka S & Toyokuni T (1996) Sphingosine transfer in cell-to-cell interaction. Biochem Biophys Res Commun, **218**: 872-876.

Jackson LS, Hlywka JJ, Senthil KR, Bullerman LB, & Musser SM (1996a) Effects of time, temperature, and pH on the stability of fumonisin B_1 in an aqueous model system. J Agric Food Chem, **44**: 906-912.

Jackson LS, Hlywka JJ, Senthil KR, & Bullerman LB (1996b) Effects of thermal processing on the stability of fumonisin B_2 in an aqueous system. J Agric Food Chem, **44**: 1984-1987.

Jackson LS, Katta SK, Fingerhut DD, DeVries JW, & Bullerman LB (1997) Effects of baking and frying on the fumonisin B_1 content of corn-based foods. J Agric Food Chem, **45**: 4800-4805.

Jaffrezou J-P, Levade T, Bettaieb A, Andrieu N, Bezombes C, Maestre N, Vermeersch S, Rousse A, & Laurent G (1996) Daunorubicin-induced apoptosis: Triggering of ceramide generation through sphingomyelin hydrolysis. EMBO J, **15**: 2417-2424.

Jaskiewicz K, Marasas WFO, & Taljaard JJF (1987a) Hepatitis in vervet monkeys caused by *Fusarium moniliforme*. J Comp Pathol, **97**: 281-291.

Jaskiewicz K, Van Rensburg SJ, Marasas WF, & Gelderblom WC (1987b) Carcinogenicity of *Fusarium moniliforme* culture material in rats. J Natl Cancer Inst, **78**: 321-325.

Jaskiewicz K, Marasas WFO, & Van der Walt FE (1987c) Oesophageal and other main cancer patterns in four districts of Transkei, 1981-1984. S Afr Med J, **72**: 27-30.

Javed T, Bennett GA, Richard JL, Dombrink-Kurtzman MA, Côté LM, & Buck WB (1993a) Mortality in broiler chicks on feed amended with a *Fusarium proliferatum* culture material or with purified fumonisin B_1 and moniliformin. Mycopathologia, **123**: 171-184.

Javed T, Richard JL, Bennett GA, Dombrink-Kurtzman MA, Bunte RM, Koelkebeck KW, Côté LM, Leeper RW, & Buck WB (1993b) Embryopathic and embryocidal effects of purified fumonisin B_1 or *Fusarium proliferatum* culture material extract on chicken embryos. Mycopathologia, **123**: 185-193.

Javed T, Dombrink-Kurtzman MA, Richard JL, Bennett GA, Côté LM, & Buck WB (1995) Sero-hematologic alterations in broiler chicks on feed amended with *Fusarium proliferatum* culture material or fumonisin B_1 and moniliformin. J Vet Diagn Invest, **7**: 520-526.

Jayadev S, Liu B, Bielawska AE, Lee JY, Nazaire F, Pushkareva MY, Obeid LM, & Hannun YA (1995) Role for ceramide in cell cycle arrest. J Biol Chem, **270**: 2047-2052.

Jeschke N, Nelson PE, & Marasas WFO (1987) Toxicity to ducklings of *Fusarium moniliforme* isolated from corn intended for use in poultry feed. Poult Sci, **66**: 1619-1623.

Johnson P, Smith EE, Phillips TD, Gentles AB, Small M, & Duffus E (1993) The effects of fumonisin B_1 on rat embryos in culture. Toxicologist, **13**: 257.

Kaneshiro T, Vesonder RF, & Peterson RE (1992) Fumonisin-stimulated N-acetyldihydro-sphingosine, *N*-acetylphytosphingosine, and phytosphingosine products of *Pichia (Hansenula) ciferri*, NRRL Y-1031. Curr Microbiol, **24**: 319-324.

Kaneshiro T, Vesonder RF, Peterson RE, & Bagby MO (1993) 2-hydroyhexadecanoic and 8,9,13-trihydroxydocosanoic acid accumulation by yeasts treated by fumonisins B_1. Lipids, **28**: 397-401.

Kang YK & Alexander JM (1996) Alterations of the glutathione redox cycle status in fumonisin B_1-treated pig kidney cells. J Biochem Toxicol, **11**: 121-126.

Kang HJ, Kim JC, Seo JA, Lee YW, & Son DH (1994) Contamination of *Fusarium* mycotoxins in corn samples imported from China. Agric Chem Biotechnol, **37**: 385-391.

Karlsson K-A (1970) Sphingolipid long chain bases. Lipids, **5**: 878-891.

Kellerman TS, Marasas WFO, Pienaar JG, & Naudé TW (1972) A mycotoxicosis of Equidae caused by *Fusarium moniliforme* Sheldon. Onderstepoort J Vet Res, **39**: 205-208.

Kellerman TS, Marasas WFO, Thiel PG, Gelderblom WCA, Cawood M, & Coetzer JAW (1990) Leukoencephalomalacia in two horses induced by oral dosing of fumonisin B_1. Onderstepoort J Vet Res, **57**: 269-275.

King SB & Scott GE (1981) Genotypic differences in maize to kernel infection by *Fusarium moniliforme*. Phytopathology, **71**: 1245-1247.

Klittich CJR, Leslie JF, Nelson PE, & Marasas WFO (1997) *Fusarium thapsinum* (*Gibberella thapsina*): A new species in section *Liseola* from sorghum. Mycologia, **89**: 643-652.

Knasmüller S, Bresgen N, Kassie F, Mersch-Sundermann V, Gelderblom W, Zöhrer E, & Eckl PM (1997) Genotoxic effects of three *Fusarium* mycotoxins, fumonisin B_1, moniliformin and vomitoxin in bacteria and in primary cultures of rat hepatocytes. Mutat Res, **391**: 39-48.

Kolesnick RN & Krönke M (1998) Regulation of ceramide production and apoptosis. Annu Rev Physiol, **60**: 643-665.

Kriek NPJ, Marasas WFO, Steyn PS, van Rensburg SL, & Steyn M (1977) Toxicity of a moniliformin-producing strain of *Fusarium moniliforme* var. *subglutinans* isolated from maize. Food Cosmet Toxicol, **15**: 579-587.

Kriek NPJ, Kellerman TS, & Marasas WFO (1981) A comparative study of the toxicity of *Fusarium verticillioides* (= *F. moniliforme*) to horses, primates, pigs, sheep and rats. Onderstepoort J Vet Res, **48**: 129-131.

Kubena LF, Edrington TS, Kamps-Holtzapple C, Harvey RB, Elissalde MH, & Rottinghaus GE (1995a) Influence of fumonisin B_1, present in *Fusarium moniliforme* culture material, and T-2 toxin on turkey poults. Poult Sci, **74**: 306-313.

Kubena LF, Edrington TS, Kamps-Holtzapple C, Harvey RB, Elissalde MH, & Rottinghaus GE (1995b) Effects of feeding fumonisin B_1 present in *Fusarium moniliforme* culture material and aflatoxin singly and in combination to turkey poults. Poult Sci, **74**: 1295-1303.

Kubena LF, Edrington TS, Harvey RB, Phillips TD, Sarr AB, & Rottinghaus GE (1997a) Individual and combined effects of fumonisin B_1 present in *Fusarium moniliforme* culture material and diacetoxyscirpenol or ochratoxin A in turkey poults. Poult Sci, **76**: 256-264.

Kubena LF, Edrington TS, Harvey RB, Buckley SA, Phillips TD, Rottinghaus GE, & Casper HH (1997b) Individual and combined effects of fumonisin B_1 present in *Fusarium moniliforme* culture material and T-2 toxin or deoxynivalenol in broiler chicks. Poult Sci, **76**: 1239-1247.

Kuiper-Goodman T, Scott PM, McEwen NP, Lombaert GA, & Ng W (1996) Approaches to the risk assessment of fumonisins in corn-based foods in Canada. Adv Exp Med Biol, **392**: 369-393.

Kwon OS, Sandberg JA, & Slikker W Jr (1997a) Effects of fumonisin B₁ treatment on blood-brain barrier transfer in developing rats. Neurotoxicol Teratol, **19**: 151-155.

Kwon OS, Schmued LC, & Slikker W Jr (1997b) Fumonisin B₁ in developing rats alters brain sphinganine levels and myelination. Neurotoxicology, **18**: 571-579.

LaBorde JB, Terry KK, Howard PC, Chen JJ, Collins FX, Shackelford ME, & Hansen DK (1997) Lack of embryotoxicity of fumonisin B₁ in New Zealand white rabbits. Fundam Appl Toxicol, **40**: 120-128.

Lamprecht SC, Marasas WFO, Alberts JF, Cawood ME, Gelderblom WCA, Shephard GS, Thiel PG, & Calitz FJ (1994) Phytotoxicity of fumonisins and TA-toxin to corn and tomato. Phytopathology, **84**: 383-391.

Laurent D, Platzer N, Kohler F, Sauviat MP, & Pellegrin F (1989a) Macrofusine et micromoniline: Deux nouvelles mycotoxines isolées de maïs infesté par *Fusarium moniliforme* Sheld. Microbiol Alim Nutr, **7**: 9-16.

Laurent D, Pellegrin F, Kohler F, Costa R, Thevenon J, Lambert C, & Huerre M (1989b) La fumonisine B₁ dans la pathogénie de la leucoencéphalomalacie équine. Microbiol Alim Nutr, **7**: 285-291.

Lavie Y, Cao H-t, Bursten SL, Guiliano AE, & Cabot MC (1996) Accumulation of glucosylceramides in multidrug-resistant cancer cells. J Biol Chem, **271**: 19530-19536.

Le Bars J, Le Bars P, Dupuy J, Boudra H, & Cassini R (1994) Biotic and abiotic factors in fumonisin B₁ production and stability. J Assoc Off Anal Chem Int, **77**: 517-521.

Lebepe-Mazur S, Bal H, Hopmans E, Murphy P, & Hendrich S (1995a) Fumonisin B₁ is fetotoxic in rats. Vet Hum Toxicol, **37**: 126-130.

Lebepe-Mazur S, Hopmans E, Murphy PA, & Hendrich S (1995b) Fed before diethyl-nitrosamine, *Fusarium moniliforme* and *F. proliferatum* mycotoxins alter the persistence of placental glutathione S-transferase-positive hepatocytes in rats. Vet Hum Toxicol, **37**: 209-214.

Lebepe-Mazur S, Wilson T, & Hendrich S (1995c) *Fusarium proliferatum*-fermented corn stimulates development of placental glutathione S-transferase-positive altered hepatic foci in female rats. Vet Hum Toxicol, **37**: 55-59.

Ledoux DR, Brown TP, Weibking TS, & Rottinghaus GE (1992) Fumonisin toxicity in broiler chicks. J Vet Diagn Invest, **4**: 330-333.

Ledoux DR, Bermudez AJ, & Rottinghaus GE (1996) Effects of feeding *Fusarium moniliforme* culture material, containing known levels of fumonisin B₁, in the young turkey poult. Poult Sci, **75**: 1472-1478.

Lee U-S, Lee M-Y, Shin K-S, Min Y-S, Cho C-M, & Ueno Y (1994) Production of fumonisin B₁ and B₂ by *Fusarium moniliforme* isolated from Korean corn kernels for feed. Mycotoxin Res, **10**: 67-72.

Lee JY, Leonhardt LG, & Obeid LM (1998) Cell-cycle-dependent changes in ceramide levels preceding retinoblastoma protein dephosphorylation in G_2/M. Biochem J, **334**: 457-461.

Lemmer ER, Hall P De La M, Gelderblom WCA, & Marasas WFO (1998) Poor reporting of oocyte apoptosis. Nat Med, **4**: 373.

Leslie JF, Plattner RD, Desjardins AE, & Klittich CJR (1992) Fumonisin B_1 production by strains from different mating populations of Gibberella fujikuroi (Fusarium section Liseola). Phytopathology, **82**: 341-345.

Leslie JF, Marasas WFO, Shephard GS, Sydenham EW, Stockenström S, & Thiel PG (1996) Duckling toxicity and the production of fumonisin and moniliformin by isolates in the A and F mating populations of Gibberella fujikuroi (Fusarium moniliforme). Appl Environ Microbiol, **62**: 1182-1187.

Lew H, Adler A, & Edinger W (1991) Moniliformin and the European corn borer. Mycotoxin Res, **7**: 71-76.

Liao W-C, Haimovitz-Friedman A, Persaud RS, McLoughlin M, Ehleiter D, Zhang N, Gatei M, Lavin M, Kolesnick R, & Fuks Z (1999) Ataxia telangiectasia-mutated gene product inhibits DNA damage-induced apoptosis via ceramide synthase. J Biol Chem, **274**: 17908-17917.

Lim CW, Parker HM, Vesonder RF, & Haschek WM (1996) Intravenous fumonisin B_1 induces cell proliferation and apoptosis in the rat. Nat Toxins, **4**: 34-41.

Logrieco A, Moretti A, Ritieni A, Chelkowski J, Altomare C, Bottalico A, & Randazzo G (1993) Natural occurrence of beauvericin in preharvest Fusarium subglutinans infected corn ears in Poland. J Agric Food Chem, **41**: 2149-2152.

Logrieco A, Moretti A, Ritieni A, Bottalico A, & Corda P (1995) Occurrence and toxigenicity of Fusarium proliferatum from preharvest maize ear rot, and associated mycotoxins, in Italy. Plant Dis, **79**: 727-731.

Logrieco A, Doko MB, Moretti A, Frisullo S, & Visconti A (1998) Occurrence of fumonisin B_1 and B_2 in Fusarium proliferatum infected asparagus plants. J Agric Food Chem, **46**: 5201-5204.

Lu Z, Dantzer WR, Hopmans EC, Prisk V, Cunnick JE, Murphy PA, & Hendrich S (1997) Reaction with fructose detoxifies fumonisin B_1 while stimulating liver-associated natural killer cell activity in rats. J Agric Food Chem, **45**: 803-809.

Lukacs Z, Schaper S, Herderich M, Schreier P, & Humpf H-U (1996) Identification and determination of fumonisin B_1 and FB_2 in corn products by high-performance liquid chromatographic-electrospray-ionization tandem mass spectrometry (HPLC-ESI-MS-MS). Chromatographia, **43**: 124-128.

Makaula NA, Marasas WFO, Venter FS, Badenhorst CJ, Bradshaw D, & Swanevelder S (1996) Oesophageal and other cancer patterns in four selected districts of Transkei, Southern Africa: 1985-1990. Afr J Health Sci, **3**:11-15.

Maragos CM (1995) Capillary zone electrophoresis and HPLC for the analysis of fluorescein isothiocyanate-labelled fumonisin B₁. J Agric Food Chem, **43**: 390-394.

Maragos CM & Richard JL (1994) Quantitation and stability of fumonisins B₁ and B₂ in milk. J Assoc Off Anal Chem Int, **77**: 1162-1167.

Marasas WFO (1993) Occurrence of *Fusarium moniliforme* and fumonisins in maize in relation to human health (Editorial). S Afr Med J, **83**: 382-383.

Marasas WFO (1994) *Fusarium*. In: Hui YM, Gorham JR, Murrell KD, & Cliver DO ed. Foodborne disease handbook. New York, Marcel Dekker, vol 2, pp 521-573.

Marasas WFO (1995) Fumonisins: Their implications for human and animal health. Nat Toxins, **3**: 193-198.

Marasas WFO (1996) Fumonisins: History, world-wide occurrence and impact. Adv Exp Med Biol, **392**: 1-17.

Marasas WFO (1997) Risk assessment of fumonisins produced by *Fusarium moniliforme* in corn. Cereal Res Commun **25**: 399-406.

Marasas WFO, Kellerman TS, Pienaar JG, & Naudé TW (1976) Leukoencephalomalacia: A mycotoxicosis of Equidae caused by *Fusarium moniliforme* Sheldon. Onderstepoort J Vet Res, **43**: 113-122.

Marasas WFO, Kriek NPJ, Wiggins VM, Steyn PS, Towers DK, & Hastie TJ (1979a) Incidence, geographic distribution and toxigenicity of *Fusarium* species in South African corn. Phytopathology, **69**: 1181-1185.

Marasas WFO, Van Rensburg SJ, & Mirocha CJ (1979b) Incidence of *Fusarium* species and the mycotoxins, deoxynivalenol and zearalenone, in corn produced in esophageal cancer areas in Transkei. J Agric Food Chem, **27**: 1108-1112.

Marasas WFO, Wehner FC, Van Rensburg SJ, & Van Schalkwyk DJ (1981) Mycoflora of corn produced in human esophageal cancer areas in Transkei, Southern Africa. Phytopathology, **71**: 792-796.

Marasas WFO, Nelson PE, & Toussoun TA (1984a) Toxigenic *Fusarium* species: Identity and mycotoxicology. University Park, Pennsylvania, The Pennsylvania State University Press, 328 pp.

Marasas WFO, Kriek NPJ, Fincham JE, & Van Rensburg SJ (1984b) Primary liver cancer and oesophageal basal cell hyperplasia in rats caused by *Fusarium moniliforme*. Int J Cancer, **34**: 383-387.

Marasas WFO, Kellerman TS, Gelderblom WCA, Coetzer JAW, Thiel PG, & Van der Lugt JJ (1988a) Leukoencephalomalacia in a horse induced by fumonisin B₁ isolated from *Fusarium moniliforme*. Onderstepoort J Vet Res, **55**: 197-203.

Marasas WFO, Jaskiewicz K, Venter FS, & Van Schalkwyk DJ (1988b) *Fusarium moniliforme* contamination of maize in oesophageal cancer areas in Transkei. S Afr Med J, **74**: 110-114.

Marasas WFO, Thiel PG, Gelderblom WCA, Shephard GS, Sydenham EW, & Rheeder JP (1993) Fumonisins produced by *Fusarium moniliforme* in maize: Foodborne carcinogens of Pan African importance. Afr Newslett Occup Health Saf, **2/93**(suppl): 11-18.

Marijanovic DR, Holt P, Norred WP, Bacon CW, Voss KA, Stancel PC, & Ragland WL (1991) Immunosuppressive effects of *Fusarium moniliforme* corn cultures in chickens. Poult Sci, **70**: 1895-1901.

Martinez-Larrañaga MR, Anadón A, Diaz MJ, Fernandez R, Sevil B, Fernandez-Cruz ML, Fernandez MC, Martinez MA, & Anton R (1996) Induction of cytochrome P4501A1 and P4504A1 activities and peroxisomal proliferation by fumonisin B_1. Toxicol Appl Pharmacol, **141**: 185-194.

Martinova EA & Merrill AH Jr (1995) Fumonisin B_1 alters sphingolipid metabolism and immune function in BALB/c mice: Immunological responses to fumonisin B_1. Mycopathologia, **130**: 163-170.

Matteri RL, Becker BA, Pace L, & Rottinghaus GE (1994) Effects of thermal environment and maternal exposure to fumonisin B_1 on luteinizing hormone secretion in preweaning gilts. Biol Reprod, **50** (suppl 1): 68.

Medlock KA & Merrill AH Jr (1988) Inhibition of serine palmitoyltransferase *in vitro* and long-chain base biosynthesis in intact Chinese hamster ovary cells by beta-chloroalanine. Biochemistry, **27**: 7079-7084.

Mehta R, Lok E, Rowsell PR, Miller JD, Suzuki CA, & Bondy GS (1998) Glutathione S-transferase-placental form expression and proliferation of hepatocytes in fumonisin B_1-treated male and female Sprague-Dawley rats. Cancer Lett, **128**: 31-39.

Meister U, Symmank H, & Dahlke H (1996) [Investigation and evaluation of the contamination of native and imported cereals with fumonisins.] Z Lebensm Unters Forsch, **203**: 528-533 (in German).

Meister U (1998) [Effect of extraction and extract clean-up on the determinable levels of fumonisins in maize and maize products: Investigations on the acid extraction and the use of immunoaffinity columns.] In: Wolff J & Betsche T ed. [Proceedings 20. Workshop on Mycotoxins, Detmold, 8-10 June 1998.] Institute of Biochemistry of Grain and Potatoes, Federal Agency for Grain, Potatoes and Fat Research, pp 241-245 (in German).

Meivar-Levy I, Sabanay H, Bershadsky AD, & Futerman AH (1997) The role of sphingolipids in the maintenance of fibroblast morphology. J Biol Chem, **272**: 1558-1564.

Meredith FI, Riley RT, Bacon CW, Williams DE, & Carlson DB (1998) Extraction, quantification, and biological availability of fumonisin B_1 incorporated into Oregon test diet and fed to rainbow trout. J Food Prot, **61**: 1034-1038.

Merrill AH Jr (1983) Characterization of serine palmitoyltransferase activity in Chinese hamster ovary cells. Biochim Biophys Acta, **754**: 284-291.

Merrill AH Jr (1991) Cell regulation by sphingosine and more complex sphingolipids. J Bioenerget Biomembr, **23**: 83-104.

Merrill AH Jr & Jones DD (1990) An update of the enzymology and regulation of sphingomyelin metabolism. Biochim Biophys Acta, **1044**: 1-12.

Merrill AH Jr, Wang E, Mullins RE, Jamison WCL, Nimkar S, & Liotta DC (1988) Quantitation of free sphingosine in liver by high-performance liquid chromatography. Anal Biochem, **171**: 373-381.

Merrill AH Jr, Hannun YA, & Bell RM (1993a) Sphingolipids and their metabolites in cell regulation. In: Bell RM, Hannun YA, & Merrill AH Jr ed. Advances in lipid research: Sphingolipids and their metabolites. Orlando, Florida, Academic Press, vol 25, pp 1-24.

Merrill AH Jr, van Echten G, Wang E, & Sandhoff K (1993b) Fumonisin B₁ inhibits sphingosine (sphinganine) *N*-acetyltransferase and *de novo* sphingolipid biosynthesis in cultured neurons *in situ*. J Biol Chem, **268**: 27299-27306.

Merrill AH Jr, Wang E, Gilchrist DG, & Riley RT (1993c) Fumonisins and other inhibitors of *de novo* sphingolipid biosynthesis. In: Bell RM, Hannun YA, & Merrill AH Jr ed. Advances in lipid research: Sphingolipids and their metabolites. Orlando, Florida, Academic Press, vol 26, pp 215-234.

Merrill AH Jr, Wang E, Schroeder JJ, Smith ER, Yoo H-S, & Riley RT (1995) Disruption of sphingolipid metabolism in the toxicity and carcinogenicity of fumonisins. In: Eklund M, Richard J, & Mise K ed. Molecular approaches to food safety issues involving toxic microorganisms. Fort Collins, Colorado, Alaken Press, pp 429-443.

Merrill AH Jr, Liotta DC, & Riley RT (1996a) Fumonisins: fungal toxins that shed light on sphingolipid function. Trends Cell Biol, **6**: 218-224.

Merrill AH Jr, Wang E, Vales TR, Smith ER, Schroeder JJ, Menaldino DS, Alexander C, Crane HM, Xia J, Liotta DC, Meredith FI, & Riley RT (1996b) Fumonisin toxicity and sphingolipid biosynthesis. Adv Exp Med Biol, **392**: 297-306.

Merrill AH Jr, Schmelz E-M, Dillehay DL, Spiegel S, Shayman JA, Schroeder JJ, Riley RT, Voss KA, & Wang E (1997a) Sphingolipids – the enigmatic lipid class: Biochemistry, physiology, and pathophysiology. Toxicol Appl Pharmacol, **142**: 208-225.

Merrill AH Jr, Schmelz E-M, Wang E, Dillehay DL, Rice LG, Meredith FI, & Riley RT (1997b) Importance of sphingolipids and inhibitors of sphingolipid metabolism as components of animal diets. J Nutr, **127**: 830S-833S.

Michel C, Van Echten-Deckert G, Rother J, Sandoff K, Wang E, & Merrill AH Jr (1997) Characterization of ceramide synthesis. J Biol Chem, **272**: 22432-22437.

Miller JD (1995) Fungi and mycotoxins in grain: Implications for stored product research. J Stored Prod Res, **31**: 1-16.

Miller JD, Savard ME, Sibilia A, Rapior S, Hocking AD, & Pitt JI (1993) Production of fumonisins and fusarins by *Fusarium moniliforme* from southeast Asia. Mycologia, **85**: 385-391.

Miller JD, Savard ME, & Rapior S (1994) Production and purification of fumonisins from a stirred jar fermenter. Nat Toxins, **2**: 354-359.

Miller JD, Savard ME, Schaafsma AW, Seifert KA, & Reid LM (1995) Mycotoxin production by *Fusarium moniliforme* and *Fusarium proliferatum* from Ontario and occurrence of fumonisin in the 1993 corn crop. Can J Plant Pathol, **17**: 233-239.

Miller MA, Honstead JP, & Lovell RA (1996) Regulatory aspects of fumonisins with respect to animal feed. Adv Exp Med Biol, **392**: 363-368.

Minervini F, Bottalico C, Pestka J, & Visconti A (1992) [On the occurrence of fumonisins in feeds in Italy.] In [Proceedings of the 46th National Congress of the Italian Society of Veterinary Science, Venice, Italy, 30 September-3 October], pp 1365-1368 (in Italian).

Mirocha CJ, Gilchrist DG, Shier WT, Abbas HK, Wen Y, & Vesonder RF (1992) AAL toxins, fumonisins (biology and chemistry) and host-specificity concepts. Mycopathologia, **17**: 47-56.

Morgan MK, Schroeder JJ, Rottinghaus GE, Powell DC, Bursian SJ, & Aulerich RJ (1997) Dietary fumonisins disrupt sphingolipid metabolism in mink and increase the free sphinganine to sphingosine ratio in urine but not in hair. Vet Hum Toxicol, **39**: 334-336.

Motelin GK, Haschek WM, Ness DK, Hall WF, Harlin KS, Schaeffer DJ, & Beasley VR (1994) Temporal and dose-response features in swine fed corn screenings contaminated with fumonisin mycotoxins. Mycopathologia, **126**: 27-40.

Mullett W, Lai EP, & Yeung JM (1998) Immunoassay of fumonisins by a surface plasmon resonance biosensor. Anal Biochem, **258**: 161-167.

Murphy PA, Rice LG, & Ross PF (1993) Fumonisin B_1, B_2, and B_3 content of Iowa, Wisconsin and Illinois corn and corn screenings. J Agric Food Chem, **41**: 263-266.

Murphy PA, Hopmans EC, Miller K, & Hendrich S (1995) Can fumonisins in foods be detoxified? In: Toxicants in food. Lancaster, Pennsylvania, Technomic Publishing, vol 1, pp 105-117.

Murphy PA, Hendrich S, Hopmans EC, Hauck CC, Lu Z, Buseman G, & Munkvold G (1996) Effect of processing on fumonisin content of corn. Adv Exp Med Biol, 392: 323-334

Musser SM & Plattner RD (1997) Fumonisin composition in cultures of *Fusarium moniliforme*, *Fusarium proliferatum*, and *Fusarium nygami*. J Agric Food Chem, **45**: 1169-1173.

Nakamura S, Kozutsumi Y, Sun Y, Miyake Y, Fujita T, & Kawasaki T (1996) Dual role of sphingolipids in signaling of the escape from and onset of apoptosis in a mouse cytotoxic T-cell line, CTLL-2. J Biol Chem, **271**: 1255-1257.

National Veterinary Services Laboratory (1995) Toxicity of fumonisins in horses (equine leukoencephalomalacia [ELEM]) - Developmental project final report. US Animal and Plant Health Inspection Agency, National Veterinary Services Laboratory (Report No. PL/TC/94-2).

Nelson PE, Plattner RD, Shackelford DD, & Desjardins AE (1991) Production of fumonisins by *Fusarium moniliforme* strains from various substrates and geographic areas. Appl Environ Microbiol, **57**: 2410-2412.

Nelson PE, Plattner RD, Shackelford DD, & Desjardins AE (1992) Fumonisin B₁ production by *Fusarium* species other than *F. moniliforme* in section *Liseola* and by some related species. Appl Environ Microbiol, **58**: 984-989.

Nelson PE, Desjardins AE, & Plattner RD (1993) Fumonisins, mycotoxins produced by *Fusarium* species: Biology, chemistry, and significance. Ann Rev Phytopathol, **31**: 233-252.

Newkirk DK, Benson RW, Howard PC, Churchwell MI, Doerge DR, & Roberts DW (1998) On-line immunoaffinity capture, coupled with HPLC and electrospray ionization mass spectrometry, for automated determination of fumonisins. J Agric Food Chem, **46**: 1677-1688.

Nikolova-Karakashian M, Morgan ET, Alexander C, Liotta DC, & Merrill AH Jr (1997) Bimodal regulation of ceramidase by interleukin-1β. J Biol Chem, **272**: 18718-18724.

Norred WP, Voss KA, Bacon CW, & Riley RT (1991) Effectiveness of ammonia treatment in detoxification of fumonisin-contaminated corn. Food Chem Toxicol, **29**: 815-819.

Norred WP, Plattner RD, Vesonder RF, Bacon CW, & Voss KA (1992a) Effects of selected secondary metabolites of *Fusarium moniliforme* on unscheduled synthesis of DNA by rat primary hepatocytes. Food Chem Toxicol, **30**: 233-237.

Norred WP, Wang E, Yoo H, Riley RT, & Merrill AH Jr (1992b) *In vitro* toxicology of fumonisins and the mechanistic implications. Mycopathologia, **117**: 73-78.

Norred WP, Plattner RD, & Chamberlain WJ (1993) Distribution and excretion of [¹⁴C]fumonisin B₁ in male Sprague-Dawley rats. Nat Toxins, **1**: 341-346.

Norred WP, Riley RT, Meredith FI, Bacon CW, & Voss KA (1996) Time- and dose-response effects of the mycotoxin, fumonisin B₁ on sphingoid base elevation in precision-cut rat liver and kidney slices. Toxicol In Vitro, **10**: 349-358.

Norred WP, Plattner RD, Dombrink-Kurtzman MA, Meredith FI, & Riley RT (1997) Mycotoxin-induced elevation of free sphingoid bases in precision-cut rat liver slices: specificity of the response and structure-activity relationships. Toxicol Appl Pharmacol, **147**: 63-70.

Ochor TE, Trevathan LE, & King SB (1987) Relationship of harvest date and host genotype to infection of maize kernels by *Fusarium moniliforme*. Plant Dis, **71**: 311-313.

Odvody GN, Remmers JC, & Spencer NM (1990) Association of kernel splitting with kernel and ear rots of corn in a commercial hybrid grown in the coastal bend of Texas. Phytopathology, **80**: 1045.

Ostrý V & Ruprich J (1998) Determination of the mycotoxin fumonisins in gluten-free diet (corn-based commodities) in the Czech Republic. Cent Eur J Public Health, **6**: 57-60.

Osweiler GD, Ross PF, Wilson TM, Nelson PE, Witte ST, Carson TL, Rice LG, & Nelson HA (1992) Characterization of an epizootic of pulmonary edema in swine associated with fumonisin in corn screenings. J Vet Diagn Invest, **4**: 53-59.

Osweiler GD, Kehrli ME, Stabel JR, Thruston JR, Ross PF, & Wilson TM (1993) Effects of fumonisin-contaminated corn screenings on growth and health of feeder calves. J Anim Sci, **71**: 459-466.

Park DL, Rua SM Jr, Mirocha CJ, Abd-Alla EAM, & Weng CY (1992) Mutagenic potentials of fumonisin contaminated corn following ammonia decontamination procedure. Mycopathologia, 117: 105-108.

Pascale M, Doko MB, & Visconti A (1995) [Determination of fumonisins in polenta by high performance liquid chromatography.] In: [Proceedings of the 2nd National Congress on Food Chemistry, Giardini-Naxos, 24-27 May 1995.] Messina, Italy, La Grafica Editoriale, pp 1067-1071 (in Italian).

Pascale M, Visconti A, Profczuk M, Wiśniewska H, & Chelkowki J (1997) Accumulation of fumonisins in maize hybrids inoculated under field conditions with *Fusarium moniliforme* Sheldon. J Sci Food Agric, 74: 1-6.

Patel S, Hazel CM, Winterton AGM, & Mortby E (1996) Survey of ethnic foods for mycotoxins. Food Addit Contam, 13: 833-841.

Patel S, Hazel CM, Winterton AGM, & Gleadle AE (1997) Surveillance of fumonisins in UK maize-based foods and other cereals. Food Addit Contam, 14: 187-191.

Paumen MB, Ishida Y, Muramatsu M, Yamamoto M, & Honjo T (1997) Inhibition of carnitine palmitoyltransferase I augments sphingolipid synthesis and palmitate-induced apoptosis. J Biol Chem, 272: 3324-3329.

Penner JD, Casteel SW, Pittman L Jr, Rottinghaus GE, & Wyatt RD (1998) Developmental toxicity of purified fumonisin B_1 in pregnant Syrian hamsters. J Appl Toxicol, 18 :197-203.

Perez GI, Knudson CM, Leykin L, Korseyer SJ, & Tilley JL (1997) Apoptosis-associated signaling pathways are required for chemotherapy-mediated female germ cell destruction. Nat Med, 3: 1228-1232.

Perry DK & Hannun YA (1998) The role of ceramide in cell signaling. Biochi Biophys Acta, 1436: 233-243.

Pestka JJ, Azcona-Olivera JI, Plattner RD, Minervini F, Doko MB, & Visconti A (1994) Comparative assessment of fumonisin in grain-based foods by ELISA, GC-MS and HPLC. J Food Prot, 57: 169-172.

Piñeiro MS, Silva GE, Scott PM, Lawrence GA, & Stack ME (1997) Fumonisin levels in Uruguayan corn products. J Assoc Off Anal Chem Int, 80: 825-828.

Pitt JI, Hocking AD, Bhudhasamai K, Miscambe BF, Wheeler KA, & Tanboon-Ek P (1993) The normal mycoflora of commodities from Thailand: 1. Nuts and oilseeds. Int J Food Microbiol, 20: 211-226.

Pittet A, Parisod V, & Schellenberg M (1992) Occurrence of fumonisins B_1 and B_2 in corn-based products from the Swiss market. J Agric Food Chem, 40: 1352-1354.

Plattner RD (1995) Detection of fumonisins produced in *Fusarium moniliforme* cultures by HPLC with electrospray MS and evaporative light scattering detectors. Nat Toxins, 3: 294-298.

Plattner RD & Branham BE (1994) Labeled fumonisins: Production and use of fumonisin B₁ containing stable isotopes. J Assoc Off Anal Chem Int, **77**: 525-532.

Plattner RD & Nelson PE (1994) Production of beauvericin by a strain of *Fusarium proliferatum* isolated from corn fodder for swine. Appl Environ Microbiol, **60**: 3894-3896.

Plattner RD & Shackelford DD (1992) Biosynthesis of labeled fumonisins in liquid cultures of *Fusarium moniliforme*. Mycopathologia, **117**: 17-22.

Plattner RD, Norred WP, Bacon CW, Voss KW, Peterspn R, Shackelford DD, & Weisleder D (1990) A method of detection of fumonisins in corn samples associated with field cases of equine leukoencephalomalacia. Mycologia, **82**: 698-702.

Plattner RD, Weisleder D, Shackelford DD, Peterson R, & Powell RG (1992) A new fumonisin from solid cultures of *Fusarium moniliforme*. Mycopathologia, **117**: 23-28.

Powell DC, Bursian SJ, Bush CR, Render JA, Rottinghaus GE, & Aulerich RJ (1996) Effects of dietary exposure to fumonisins from *Fusarium moniliforme* culture material (M1325) on the reproductive performance of female mink. Arch Environ Contam Toxicol, **31**: 286-292.

Prathapkumar SH, Rao VS, Paramkishan RJ, & Bhat RV (1997) Disease outbreak in laying hens arising from the consumption of fumonisin-contaminated food. Br Poult Sci, **38**: 475-479.

Prelusky DB, Trenholm, HL, & Savard ME (1994) Pharmacokinetic fate of ^{14}C-labelled fumonisin B₁ in swine. Nat Toxins, **2**: 73-80.

Prelusky DB, Savard ME, & Trenholm HL (1995) Pilot study on the plasma pharmacokinetics of fumonisin B₁ in cows following a single dose by oral gavage or intravenous administration. Nat Toxins, **3**: 389-394.

Prelusky DB, Trenholm HL, Rotter BA, Miller JD, Savard ME, Yeung JM, & Scott PM (1996a) Biological fate of fumonisin B₁ in food-producing animals. Adv Exp Med Biol, **392**: 265-278.

Prelusky DB, Miller JD, & Trenholm HL (1996b) Disposition of ^{14}C-derived residues in tissues of pigs fed radiolabelled fumonisin B₁. Food Addit Contam, **13**: 155-162.

Price WD, Lovell RA, & McChesney DG (1993) Naturally occurring toxins in feedstuffs: Center for Veterinary Medicine perspective. J Anim Sci, **71**: 2556-2562.

Qureshi MA & Hagler WM Jr (1992) Effect of fumonisin-B₁ exposure on chicken macrophage function *in vitro*. Poult Sci, **71**: 104-112.

Radin NS (1994) Glucosylceramide in the nervous system: a mini review. Neurochem Res, **19**: 533-540.

Ramasamy S, Wang E, Hennig B, & Merrill AH Jr (1995) Fumonisin B₁ alters sphingolipid metabolism and disrupts the barrier function of endothelial cells in culture. Toxicol Appl Pharmacol, **133**: 343-348.

Ramirez ML, Pascale M, Chulze S, Reynoso MM, March G, & Visconti A (1996) Natural occurrence of fumonisins and their correlation to *Fusarium* contamination in commercial corn hybrids growth in Argentina. Mycopathologia, **135**: 29-34.

Ramljak D, Diwan BA, Ramakrishna G, Victor TC, Marasas WFO, & Gelderblom WCA (1996) Overexpression of cyclin D1 is an early event in, and possible mechanism responsible for, fumonisin B$_1$ liver tumorigenesis in rats. Proc Am Assoc Cancer Res, **38**: 495.

Rapior S, Miller JD, Savard ME, & Apsimon JW (1993) Production *in vitro* de fumonisines et de fusarines par des souches européennes de *Fusarium moniliforme*. Microbiol Alim Nutr, **11**: 327-333.

Reddy RV, Johnson G, Rottinghaus GE, Casteel SW, & Reddy CS (1996) Developmental effects of fumonisin B$_1$ in mice. Mycopathologia, **134**: 161-166.

Restum JC, Bursian SJ, Millerick M, Render JA, Merrill AH Jr, Wang E, Rottinghaus GE, & Aulerich RJ (1995) Chronic toxicity of fumonisins from *Fusarium moniliforme* culture material (M-1325) to mink. Arch Environ Contam Toxicol, **29**: 545-550.

Rheeder JP, Marasas WFO, Thiel PG, Sydenham EW, Shephard GS, & van Schalkwyk DJ (1992) *Fusarium moniliforme* and fumonisins in corn in relation to human esophageal cancer in Transkei. Phytopathology, **82**: 353-357.

Rheeder JP, Marasas WFO, Farina MPW, Thompson GR, & Nelson PE (1994) Soil fertility factors in relation to oesophageal cancer risk areas in Transkei, Southern Africa. Eur J Cancer Prev, **3**: 49-56.

Rice LG & Ross PF (1994) Methods for detection and quantitation of fumonisins in corn, cereal products and animal excreta. J Food Prot, **57**: 536-540.

Richard JL, Meerdink G, Maragos CM, Tumbleson M, Bordson G, Rice LG, & Ross PF (1996) Absence of detectable fumonisins in the milk of cows fed *Fusarium proliferatum* (Matsushima) Nirenberg culture material. Mycopathologia, **133**: 123-126.

Riley RT & Yoo H-S (1995) Time and dose relationship between the cellular effects of fumonisin B$_1$ (FB$_1$) and the uptake and accumulation of [^{14}C]FB in LLC$_1$-PK cells. Toxicologist, **15**(1): 290.

Riley RT, An N-H, Showker JL, Yoo H-S, Norred WP, Chamberlain WJ, Wang E, Merrill AH Jr, Motelin G, Beasley VR, & Haschek WM (1993) Alteration of tissue and serum sphinganine to sphingosine ratio: An early biomarker of exposure to fumonisin-containing feeds in pigs. Toxicol Appl Pharmacol, **118**: 105-112.

Riley RT, Hinton DM, Chamberlain WJ, Bacon CW, Wang E, Merrill AH Jr, & Voss KA (1994a) Dietary fumonisin B$_1$ induces disruption of sphingolipid metabolism in Sprague-Dawley rats: A new mechanism of nephrotoxicity. J Nutr, **124**: 594-603.

Riley RT, Voss KA, Yoo H-S, Gelderblom WCA, & Merrill AH Jr (1994b) Mechanism of fumonisin toxicity and carcinogenesis. J Food Prot, **57**: 638-645.

Riley RT, Wang E, & Merrill AH Jr (1994c) Liquid chromatographic determination of sphinganine and sphingosine: Use of the free sphinganine-to-sphingosine ratio as a biomarker for consumption of fumonisins. J Assoc Off Anal Chem Int, **77**: 533-540.

Riley RT, Wang E, Schroeder JJ, Smith ER, Plattner RD, Abbas H, Yoo H-S, & Merrill AH Jr (1996) Evidence for disruption of sphingolipid metabolism as a contributing factor in the toxicity and carcinogenicity of fumonisins. Nat Toxins, **4**: 3-15.

Riley RT, Showker JL, Owens DL, & Ross PF (1997) Disruption of sphingolipid metabolism and induction of equine leukoencephalomalacia by *Fusarium proliferatum* culture material containing fumonisin B_1 or B_3. Environ Toxicol Pharmacol, 3: 221-228.

Riley RT, Voss KA, Norred WP, Wang E, & Merrill AH Jr (1998) Fumonisins: mechanism of mycotoxicity. Rev Méd Vét, **149**: 617-626.

Riley RT, Voss KA, Norred WP, Bacon CW, Meredith FI, & Sharma RP (1999) Serine palmitoyltransferase inhibition reverses anti-proliferative effects of ceramide synthase inhibition in cultured renal cells and suppresses free sphingoid base accumulation in kidney of BALBc mice. Environ Toxicol Pharmacol, 7: 109-118.

Ross PF (1994) What are we going to do with this dead horse? J Assoc Off Anal Chem Int, **77**: 491-494.

Ross PF, Nelson PE, Richard JL, Osweiler GD, Rice LG, Plattner RD, & Wilson TM (1990) Production of fumonisins by *Fusarium moniliforme* and *Fusarium proliferatum* isolates associated with equine leukoencephalomalacia and a pulmonary edema syndrome in swine. Appl Environ Microbiol, **56**: 3225-3226.

Ross PF, Rice LG, Plattner RD, Osweiler GD, Wilson TM, Owens DL, Nelson HA, & Richard JL (1991a) Concentrations of fumonisin B_1 in feeds associated with animal health problems. Mycopathologia, **114**: 129-135.

Ross PF, Rice LG, Reagor JC, Osweiler GD, Wilson TM, Nelson HA, Owens DL, Plattner RD, Harlin KA, Richard JL, Colvin BM, & Banton MI (1991b) Fumonisin B_1 concentrations in feeds from 45 confirmed equine leukoencephalomalacia cases. J Vet Diagn Invest, **3**: 238-241.

Ross PF, Rice LG, Osweiler GD, Nelson PE, Richard JL, & Wilson TM (1992) A review and update of animal toxicoses associated with fumonisin-contaminated feeds and production of fumonisins by *Fusarium* isolates. Mycopathologia, **117**: 109-114.

Ross PF, Ledet AE, Owens DL, Rice LG, Nelson HA, Osweiler GD, & Wilson TM (1993) Experimental equine leukoencephalomalacia, toxic hepatosis, and encephalopathy caused by corn naturally contaminated with fumonisins. J Vet Diagn Invest, **5**: 69-74.

Ross PF, Nelson PE, Owens DL, Rice LG, Nelson HA, & Wilson TM (1994) Fumonisin B_2 in cultured *Fusarium proliferatum*, M-6104, causes equine leukoencephalomalacia. J Vet Diagn Invest, **6**: 263-265.

Rother J, Van Echten G, Schwarzmann G, & Sandhoff K (1992) Biosynthesis of sphingolipids: Dihydroceramide and not sphinganine is desaturated by cultured cells. Biochem Biophys Res Commun, **189**: 14-20.

Rotter BA, & Oh Y-N (1996) Mycotoxin fumonisin B_1 stimulates nitric oxide production in a murine macrophage cell line. Nat Toxins, **4**: 291-294.

Rotter BA, Thompson BK, Prelusky DB, Trenholm HL, Stewart B, Miller JD, & Savard ME (1996) Response of growing swine to dietary exposure pure fumonisin B_1 during an eight-week period: Growth and clinical parameters. Nat Toxins, **4**: 42-50.

Rottinghaus GE, Coatney CE, & Minor HC (1992) A rapid, sensitive thin layer chromatography procedure for the detection of fumonisin B_1 and B_2. J Vet Diagn Invest, **4**: 326-329.

Rumbeiha WK & Oehme FW (1997) Fumonisin exposure to Kansans through consumption of corn-based market foods. Vet Hum Toxicol, **39**: 220-225.

Sahu SC, Eppley RM, Page SW, Gray GC, Barton CN, & O'Donnell MW (1998) Peroxidation of membrane lipids and oxidative DNA damage by fumonisin B_1 in isolated rat liver nuclei. Cancer Lett, **125**: 117-121.

Sakata K, Sakata A, Vela-Roch N, Espinosa R, Escalante A, Kong L, Nakabayashi T, Cheng J, Talal N, & Dang H (1998) Fas (CD95)-transduced signal preferentially stimulates lupus peripheral T lymphocytes. Eur J Immunol, **28**: 2648-2660.

Sanchis V, Abadias M, Oncins L, Sala N, Viñas I, & Canela R (1994) Occurrence of fumonisins B_1 and B_2 in corn-based products from the Spanish market. Appl Environ Microbiol, **60**: 2147-2148.

Sanchis V, Abadias M, Oncins L, Sala N, Viñas I, & Canela R (1995) Fumonisins B_1 and B_2 and toxigenic *Fusarium* strains in feeds from the Spanish market. Intl J Food Microbiol, **27**: 37-44.

Sandvig K, Garred O, Van Helvoort A, Van Meer G, & Van Deurs B (1996) Importance of glycolipid synthesis for butyric acid-induced sensitization to Shiga toxin and intracellular sorting of toxin in A431 cells. Mol Biol Cell, **7**: 1391-1404.

Sauviat MP, Laurent D, Kohler F, & Pellegrin F (1991) Fumonisin, a toxin from the fungus *Fusarium moniliforme* Sheld, blocks both the calcium current and the mechanical activity in frog atrial muscle. Toxicon, **29**: 1025-1031.

Savard ME & Blackwell BA (1994) Spectral characteristics of secondary metabolites from *Fusarium* fungi. In: Miller JD & Trenholm HL ed. Mycotoxins in grain: Compounds other than aflatoxin. St. Paul, Minnesota, Eagan Press, pp 59-260.

Schaafsma AW, Miller JD, Savard ME, & Ewing RJ (1993) Ear rot development and mycotoxin production in corn in relation to inoculation method, corn hybrid, and species of *Fusarium*. Can J Plant Pathol, **15**: 185-192.

Schmelz EM, Dombrink-Kurtzman MA, Roberts PC, Kozutsumi Y, Kawasaki T, & Merrill AH Jr (1998) Induction of apoptosis by fumonisin B_1 in HT29 cells is mediated by the accumulation of endogenous free sphingoid bases. Toxicol Appl Pharmacol, **148**: 252-260.

Schroeder JJ, Crane HM, Xia J, Liotta DC, & Merrill AH Jr (1994) Disruption of sphingolipid metabolism and stimulation of DNA synthesis by fumonisin B_1: A molecular mechanism for carcinogenesis associated with *Fusarium moniliforme*. J Biol Chem, **269**: 3475-3481.

Schwarz A, Rapaport E, Hirschberg K, & Futerman H (1995). A regulatory role for sphingolipids in neuronal growth: Inhibition of sphingolipid synthesis and degradation have opposite effects on axonal branching. J Biol Chem, **270**: 10990-10998.

Scott PM (1993) Fumonisins. Int J Food Microbiol, **18**: 257-270.

Scott PM & Lawrence GA (1992) Liquid chromatographic determination of fumonisins with 4-fluoro-7-nitrobenzofurazan. J Assoc Off Anal Chem Int, **75**: 829-834.

Scott PM & Lawrence GA (1994) Stability and problems in recovery of fumonisins added to corn-based foods. J Assoc Off Anal Chem Int, **77**: 541-545.

Scott PM & Lawrence GA (1995) Analysis of beer for fumonisins. J Food Prot, **58**: 1379-1382.

Scott PM & Lawrence GA (1996) Determination of hydrolysed fumonisin B₁ in alkali-processed corn foods. Food Addit Contam, **13**: 823-832.

Scott PM & Trucksess MW (1997) Application of immunoaffinity columns to mycotoxin analysis. J Assoc Off Anal Chem Int, **80**: 941-949.

Scott PM, Delgado T, Prelusky DB, Trenholm HL, & Miller JD (1994) Determination of fumonisins in milk. J Environ Sci Health, **B29**: 989-998.

Scott PM, Kanhere SR, Lawrence GA, Daley EF, & Farber JM (1995) Fermentation of wort containing added ochratoxin A and fumonisins B₁ and B₂. Food Addit Contam, **12**: 31-40.

Scott PM, Yeung JM, Lawrence GA, & Prelusky DB (1997) Evaluation of enzyme-linked immunosorbent assay for analysis of beer for fumonisins. Food Addit Contam, **14**: 445-450.

Scudamore KA & Chan HK (1993) Occurrence of fumonisin mycotoxins in maize and millet imported into the United Kingdom. In: Scudamore K ed. Occurrence and significance of mycotoxins. Slough, United Kingdom, Ministry of Agriculture, Fisheries and Food, pp 186-189.

Scudamore KA, Nawaz S, & Hetmanski MT (1998) Mycotoxins in ingredients of animal feeding stuffs: II. Determination of mycotoxins in maize and maize products. Food Addit Contam, **15**: 30-55.

Sharma RP, Dugyala RR, & Voss KA (1997) Demonstration of in-situ apoptosis in mouse liver and kidney after short-term repeated exposure to fumonisin B₁. J Comp Pathol, **117**: 371-381.

Shelby RA, Rottinghaus GE, & Minor HC (1994a) Comparison of thin-layer chromatography and competitive immunoassay methods for detecting fumonisin on maize. J Agric Food Chem, **42**: 2064-2067.

Shelby RA, White DG, & Bauske EM (1994b) Differential fumonisin production in maize hybrids. Plant Dis, **78**: 582-584.

Shephard GS (1998) Chromatographic determination of the fumonisin mycotoxins. J Chromatogr, **A815**: 31-39.

Shephard GS, Sydenham EW, Thiel PG, & Gelderblom WCA (1990) Quantitative determination of fumonisins B₁ and B₂ by high-performance liquid chromatography with fluorescence detection. J Liq Chromatogr, **13**: 2077-2087.

Shephard GS, Thiel PG, & Sydenham EW (1992a) Initial studies on the toxicokinetics of fumonisin B_1 in rats. Food Chem Toxicol, 30: 277-279.

Shephard GS, Thiel PG, Sydenham EW, Alberts JF, & Gelderblom WCA (1992b) Fate of a single dose of the ^{14}C-labelled mycotoxin fumonisin B_1, in rats. Toxicon, 30: 768-770.

Shephard GS, Thiel PG, & Sydenham EW (1992c) Determination of fumonisin B_1 in plasma and urine by high-performance liquid chromatography. J Chromatogr, 574: 299-304.

Shephard GS, Thiel PG, Sydenham EW, Alberts JF, & Cawood ME (1994a) Distribution and excretion of a single dose of the mycotoxin fumonisin B_1 in a non-human primate. Toxicon, 32: 735-741.

Shephard GS, Thiel PG, Sydenham EW, Vleggaar R, & Alberts JF (1994b) Determination of the mycotoxin fumonisin B_1 and identification of its partially hydrolysed metabolites in the faeces of non-human primates. Food Chem Toxicol, 32: 23-29.

Shephard GS, Thiel PG, Sydenham EW, & Alberts JF (1994c) Biliary excretion of the mycotoxin fumonisin B_1 in rats. Food Chem Toxicol, 32: 489-491.

Shephard GS, Thiel PG, Sydenham EW, & Savard ME (1995a) Fate of a single dose of ^{14}C-labelled fumonisin B_1 in vervet monkeys. Nat Toxins, 3: 145-150.

Shephard GS, Thiel PG, Sydenham EW, & Snijman PW (1995b) Toxicokinetics of the mycotoxin fumonisin B_2 in rats. Food Chem Toxicol, 33: 591-595.

Shephard GS, Thiel PG, & Sydenham EW (1995c) Liquid chromatographic determination of the mycotoxin fumonisin B_2 in physiological samples. J Chromatogr, A692: 39-43.

Shephard GS, Thiel PG, Stockenström S, & Sydenham EW (1996a) Worldwide survey of fumonisin contamination of corn and corn-based products. J Assoc Off Anal Chem Int, 79: 671-687.

Shephard GS, van der Westhuizen L, Thiel PG, Gelderblom WCA, Marasas WFO, & Van Schalkwyk DJ (1996b) Disruption of sphingolipid metabolism in non-human primates consuming diets of fumonisin-containing *Fusarium moniliforme* culture material. Toxicon, 34: 527-534.

Shetty PH & Bhat RV (1997) Natural occurrence of fumonisin B_1 and its co-occurrence with aflatoxin B_1 in Indian sorghum, maize and poultry feeds. J Agric Food Chem, 45: 2170-2173.

Shetty PH & Bhat RV (1998) Sensitive method for the detection of fumonisin B_1 in human urine. J Chromatogr, B705: 171-173.

Sheu CW, Rodriguez I, Eppley RM, & Lee JK (1996) Lack of transforming activity of fumonisin B_1 in BALB/3T3 A31-1-1 mouse embryo cells. Food Chem Toxicol, 34: 751-753.

Shimabukuro M, Zhou Y-T, Levi M, & Unger RH (1998) Fatty acid-induced β cell apoptosis: a link between obesity and diabetes. Proc Natl Acad Sci (USA), 95: 2498-2502.

Shurtleff MC (1980) Compendium of corn diseases. St. Paul, Minnesota, American Phytopathological Society, p 105.

Smith ER & Merrill AH Jr (1995) Differential roles of *de novo* sphingolipid biosynthesis and turnover in the "burst" of free sphingosine and sphinganine, and their 1-phosphates and N-acyl-derivatives, that occurs upon changing the medium of cells in culture. J Biol Chem, **270**: 18749-18758.

Smith JS & Thakur RA (1996) Occurrence and fate of fumonisins in beef. Adv Exp Med Biol, **392**: 39-55.

Smith GW, Constable PD, Bacon CW, Meredith FI, & Haschek WM (1996a) Cardiovascular effects of fumonisins in swine. Fundam Appl Toxicol, **31**: 169-172.

Smith GW, Constable PD, & Haschek WM (1996b) Cardiovascular responses to short-term fumonisin exposure in swine. Fundam Appl Toxicol, **33**: 140-148.

Smith GW, Constable PD, Smith AR, Bacon CW, Meredith FI, Wollenberg GK, & Haschek WM (1996c) Effects of fumonisin-containing culture material on pulmonary clearance in swine. Am J Vet Res, **57**: 1233-1238.

Smith ER, Jones PL, Boss JM, & Merrill AH Jr (1997) Changing J774A.1 cells to new medium perturbs multiple signaling pathways, including the modulation of protein kinase C by endogenous sphingoid bases. J Biol Chem, **272**: 5640-5646.

Solfrizzo M, Avantaggiato G, & Visconti A (1997a) Rapid method to determine sphinganine/sphingosine in human and animal urine as a biomarker for fumonisin exposure. J Chromatogr Biomed Appl, **692**: 87-93.

Solfrizzo M, Avantaggiato G, & Visconti A (1997b) *In vivo* validation of the sphinganine/sphingosine ratio as a biomarker to display fumonisin ingestion. Cereal Res Commun, **25**: 437-441.

Spiegel S (1999) Sphingosin 1-phosphate: a prototype of a new class of second messengers. J Leukoc Biol, **65**: 341-344.

Stack ME (1998) Analysis of fumonisin B₁ and its hydrolysis product in tortillas. J Assoc Off Anal Chem Int, **81**: 737-740.

Stack ME & Eppley RM (1992) Liquid chromatographic determination of fumonisins B₁ and B₂ in corn and corn products. J Assoc Off Anal Chem Int, **75**: 834-837.

Stevens VL & Tang J (1997) Fumonisin B₁-induced sphingolipid depletion inhibits vitamin uptake via the glycosylphosphatidylinositol-anchored folate receptor. J Biol Chem, **272**: 18020-18025.

Stevens VL, Nimkar S, Jamison WC, Liotta DC, & Merrill AH Jr (1990) Characteristics of the growth inhibition and cytotoxicity of long-chain (sphingoid) bases for Chinese hamster ovary cells: Evidence for an involvement of protein kinase C. Biochim Biophys Acta, **1051**: 37-45.

Stockenström S, Sydenham EW, & Thiel PG (1994) Determination of fumonisins in corn: Evaluation of two purification procedures. Mycotoxin Res, **10**: 9-14.

Strum JC, Swenson KI, Turner JE, & Bell RM (1995) Ceramide triggers meiotic cell cycle progression in *Xenopus* oocytes. A potential mediator of progesterone-induced maturation. J Biol Chem, **270**: 13541-13547.

Sun TSC & Stahr HM (1993) Evaluation and application of a bioluminescent bacterial genotoxicity test. J Assoc Off Anal Chem Int, 76: 893-898.

Suzuki CAM, Hierlihy L, Barker M, Curran I, Mueller R, & Bondy GS (1995) The effects of fumonisins on several markers of nephrotoxicity in rats. Toxicol Appl Pharmacol, **133**: 207-214.

Suzuki A, Iwasaki M, Kato M, & Wagai N (1997) Sequential operation of ceramide synthesis and ICE cascade in CPT-11-initiated apoptotic death signaling. Exp Cell Res, **233**: 41-47.

Sweeley CC (1991) Sphingolipids. In: Vance DE & Vance JE ed. Biochemistry of lipids, lipoproteins and membranes. Amsterdam, Elsevier Science Publishers, pp 327-361.

Sweeney EA, Sakakura C, Shirahama T, Masamune A, Ohta H, Hakomori S-I, & Igarashi Y (1996) Sphingosine and its methylated derivative N,N-methylsphingosine (DMS) induce apoptosis in a variety of human cancer cell lines. Int J Cancer, **66**: 358-366.

Swiss Federal Office of Public Health (1997a) Révision d'ordonnance sur les substances étrangères et les composants OSEC (RS 817.021.23). Bern, Swiss Federal Office of Public Health, 2 pp (Circular No. 11).

Swiss Federal Office of Public Health (1997b) Liquid chromatographic determination of fumonisins B_1, B_2, and B_3 in foods and feeds. J Assoc Off Anal Chem Int, **75**: 313-318.

Sydenham EW (1994) Fumonisins: Chromatographic methodology and their role in human and animal health. Cape Town, South Africa, University of Cape Town (Ph.D. Thesis).

Sydenham EW & Shephard GS (1996) Chromatographic and allied methods of analysis for selected mycotoxins. In: Gilbert J ed. Progress in food contaminant analysis. London, Blackie, pp 65-146.

Sydenham EW, Gelderblom WCA, Thiel PG, & Marasas WFO (1990a) Evidence for the natural occurrence of fumonisin B_1, a mycotoxin produced by *Fusarium moniliforme*, in corn. J Agric Food Chem, **38**: 285-290.

Sydenham EW, Thiel PG, Marasas WFO, Shephard GS, Van Schalkwyk DJ, & Koch KR (1990b) Natural occurrence of some *Fusarium* mycotoxins in corn from low and high esophageal cancer prevalence areas of the Transkei, Southern Africa. J Agric Food Chem, **38**: 1900-1903.

Sydenham EW, Shephard GS, Thiel PG, Marasas WFO, & Stockenström S (1991) Fumonisin contamination of commercial corn-based human foodstuffs. J Agric Food Chem, **39**: 2014-2018.

Sydenham EW, Marasas WFO, Shephard GS, Thiel PG, & Hirooka EY (1992) Fumonisin concentrations in Brazilian feeds associated with field outbreaks of confirmed and suspected animal mycotoxicoses. J Agric Food Chem, **40**: 994-997.

Sydenham EW, Shepard GS, Gelderblom WCA, Thiel PG, & Marasas WFO (1993a) Fumonisins: their implications for human and animal health. In: Scudamore K ed. Proceedings of the UK Workshop on Occurrence and Significance of Mycotoxins. Slough, United Kingdom, Ministry of Agriculture, Fisheries and Food, Central Science Laboratory, pp 42-48.

Sydenham EW, Shephard GS, Thiel PG, Marasas WFO, Rheeder JP, Sanhueza CEP, González HHL, & Resnick SL (1993b) Fumonisins in Argentinian field-trial corn. J Agric Food Chem, **41**: 891-895.

Sydenham EW, Stockenström S, Thiel PG, Shephard GS, Koch KR, & Marasas WFO (1995) Potential of alkaline hydrolysis for the removal of fumonisins from contaminated corn. J Agric Food Chem, **43**: 1198-1201.

Sydenham EW, Shephard GS, Thiel PG, Stockenström S, Snijman PW, & Van Schalkwyk DJ (1996) Liquid chromatographic determination of fumonisins B₁, B₂, and B₃ in corn: AOAC-IUPAC collaborative study. J Assoc Off Anal Chem Int, **79**: 688-696.

Sydenham EW, Shephard GS, Stockenström S, Rheeder JP, Marasas WFO, & van der Merwe MJ (1997) Production of fumonisin B analogues and related compounds by *Fusarium globosum*, a newly described species from corn. J Agric Food Chem, **45**: 4004-4010.

Tejada-Simon MV, Marovatsanga LT, & Pestka JJ (1995) Comparative detection of fumonisin by HPLC, ELISA, and immunocytochemical localization in *Fusarium* cultures. J Food Prot, **58**: 666-672.

Thiel PG, Marasas WFO, Sydenham EW, Shephard GS, Gelderblom WCA, & Nieuwenhuis JJ (1991a) Survey of fumonisin production by *Fusarium* species. Appl Environ Microbiol, **57**: 1089-1093.

Thiel PG, Shephard GS, Sydenham EW, Marasas WFO, Nelson PE, & Wilson TM (1991b) Levels of fumonisins B₁ and B₂ in feeds associated with confirmed cases of equine leukoencephalomalacia. J Agric Food Chem, **39**: 109-111.

Thiel PG, Marasas WFO, Sydenham EW, Shephard GS, & Gelderblom WCA (1992) The implications of naturally occurring levels of fumonisins in corn for human and animal health. Mycopathologia, **117**: 3-9.

Thiel PG, Sydenham EW, Shephard GS, & Van Schalkwyk DJ (1993) Study of the reproducibility characteristics of a liquid chromatographic method for the determination of fumonisins B₁ and B₂ in corn: IUPAC collaborative study. J Assoc Off Anal Chem Int, **76**: 361-366.

Tolleson WH, Dooley KL, Sheldon WG, Thurman JD, Bucci TJ, & Howard PC (1996a) The mycotoxin fumonisin induces apoptosis in cultured human cells and in livers and kidneys of rats. Adv Exp Med Biol, **392**: 237-250.

Tolleson WH, Melchior WB Jr, Morris SM, McGarrity LJ, Domon OE, Muskhelishvili L, James SJ, & Howard PC (1996b) Apoptotic and antiproliferative effects of fumonisin B₁ in human keratinocytes, fibroblasts, esophageal epithelial cells and hepatoma cells. Carcinogenesis, **17**: 239-249.

Tolleson WH, Couch LH, Melchior WB Jr, Jenkins GR, Muskhelishvili M, Muskhelishvili L, McGarrity LJ, Domon O, Morris SM, & Howard PC (1999) Fumonisin B₁ induces apoptosis in cultured human keratinocytes through sphinganine accumulation and ceramide depletion. Int J Oncol, **14**: 833-843.

Torres MR, Sanchis V, & Ramos AJ (1998) Occurrence of fumonisins in Spanish beers analyzed by an enzyme-linked immunosorbent assay method. Int J Food Microbiol, **39**: 139-143.

Trucksess MW, Stack ME, Allen S, & Barrion N (1995) Immunoaffinity column coupled with liquid chromatography for determination of fumonisin B_1 in canned and frozen sweet corn. J Assoc Off Anal Chem Int, **78**: 705-710.

Tryphonas H, Bondy G, Miller JD, Lacroix F, Hodgen M, McGuire P, Fernie S, Miller D, & Hayward S (1997) Effects of fumonisin B_1 on the immune system of Sprague-Dawley rats following a 14-day oral (gavage) exposure. Fundam Appl Toxicol, **39**: 53-59.

Tseng T-C & Liu C-Y (1997) Natural occurrence of fumonisins in corn-based foodstuffs in Taiwan. Cereal Res Commun, **25**: 393-394.

Tsunoda M, Sharma RP, & Riley RT (1998) Early fumonisin B_1 toxicity in relation to disrupted sphingolipid metabolism in male BALB/c mice. J Biochem Mol Toxicol, **12**: 281-289.

Ueda N, Kaushal GP, Hong X, & Shah SV (1998) Role of enhanced ceramide generation in DNA damage and cell death in chemical hypoxic injury to LLC-PK_1 cells. Kidney Int, **54**: 399-406.

Ueno Y, Aoyama S, Sugiura Y, Wang D-S, Lee U-S, Hirooka EY, Hara S, Karki T, Chen G, & Yu S-Z (1993) A limited survey of fumonisins in corn and corn-based products in Asian countries. Mycotoxin Res, **9**: 27-34.

Ueno Y, Umemori K, Niimi E, Tanuma S, Nagata S, Sugamata M, Ihara T, Sekijima M, Kawai K, Ueno I, & Tashiro F (1995) Induction of apoptosis by T-2 toxin and other natural toxins in HL-60 human promyelotic leukemia cells. Nat Toxins, **3**: 129-137.

Ueno Y, Iijima K, Wang S-D, Sugiura Y, Sekijima M, Tanaka T, Chen C, & Yu S-Z (1997) Fumonisins as a possible contributory risk factor for primary liver cancer: A 3-year study of corn harvested in Haimen, China by HPLC and ELISA. Food Chem Toxicol, **35**: 1143-1150.

Usleber E, Straka M, & Terplan G (1994a) Enzyme immunoassay for fumonisin B_1 applied to corn-based food. J Agric Food Chem, **42**: 1392-1396.

Usleber E, Schlichtherle C, Märtlbauer E (1994b) DMZ. Mag Food Dairy Ind, **115**: 1220-1227.

US NTP (1999) NTP technical report on the toxicology and carcinogenesis studies of fumonisin B_1 (CAS No. 116355-83-0) in F344/N rats and B6C3F_1 mice (feed studies). Research Triangle Park, North Carolina, US Department of Health and Human Services, National Toxicology Program (NTP TR 496; NIH Publication No. 99-3955).

Van Asch MAJ, Rijkenberg FHJ, & Coutinho TA (1992) Phytotoxicity of fumonisin B_1, moniliformin, and T-2 toxin to corn callus cultures. Phytopathology, **82**: 1330-1332.

Van der Westhuizen L, Shephard GS, Snyman SD, Abel S, Swanevelder S, & Gelderblom WCA (1998) Inhibition of sphingolipid biosynthesis in rat primary hepatocyte cultures by fumonisin B_1 and other structurally related compounds. Food Chem Toxicol, **36**: 497-503.

Van Veldhoven PP & Mannaerts GP (1993) Sphingosine-phosphate lyase. In: Bell RM, Hannun YA, & Merrill AH Jr ed. Advances in lipid research: Sphingolipids and their metabolites. Orlando, Florida, Academic Press, vol 26, pp 69-98.

Velázquez C, van Bloemendal C, Sanchis V, & Canela R (1995) Derivation of fumonisins B₁ and B₂ with 6-aminoquinolyl N-hydroxysuccinimidylcarbamate. J Agric Food Chem, **43**: 1535-1537.

Vesonder RF & Wu W (1998) Correlation of moniliformin, but not fumonisin B₁ levels, in culture materials of *Fusarium* isolates to acute death in ducklings. Poult Sci, **77**: 67-72.

Vesonder RF, Labeda DP, & Peterson RE (1992) Phytotoxic activity of selected water-soluble metabolites of *Fusarium* against *Lemma minor* L. (Duckweed). Mycopathologia, **118**: 185-189.

Viljoen JH, Marasas WFO, & Thiel PG (1994) Fungal infection and mycotoxic contamination of commercial maize. In: Du Plessis JG, Van Rensberg JBJ, McLaren NW, & Flett BC ed. Proceedings of the Tenth South African Maize Breeding Symposium. Potchefstroom, South Africa, Department of Agriculture, pp 26-37.

Visconti A (1996) Fumonisins in maize genotypes grown in various geographic areas. Adv Exp Med Biol, **392**: 193-204.

Visconti A & Boenke A (1995) Improvement of the determination of fumonisins (FB₁ and FB₂) in maize and maize based feeds. Brussels, European Commission, BCR Information, 101 pp (Report EUR 16617 EN).

Visconti A & Doko MB (1994) Survey of fumonisin production by *Fusarium* isolated from cereals in Europe. J Assoc Off Anal Chem Int, **77**: 546-550.

Visconti A, Doko MB, Schurer B, & Boenke A (1993) Intercomparison study for the analysis of fumonisins B₁ and B₂ in an unknown solution. In. Scudamore K ed. Proceedings of the UK Workshop on Occurrence and Significance of Mycotoxins. Slough, United Kingdom, Ministry of Agriculture, Fisheries and Food, Central Science Laboratory, pp 200-202.

Visconti A, Doko MB, Bottalico C, Schurer B, & Boenke A (1994) Stability of fumonisins (FB₁ and FB₂) in solution. Food Addit Contam, **11**: 427-431.

Visconti A, Boenke A, Doko MB, Solfrizzo M, & Pascale M (1995) Occurrence of fumonisins in Europe and the BCR-measurements and testing projects. Nat Toxins, **3**: 269-274.

Voss KA, Plattner RD, Bacon CW, & Norred WP (1990) Comparative studies of hepatotoxicity and fumonisin B₁ and B₂ content of water and chloroform/methanol extracts of *Fusarium moniliforme* strain MRC 826 culture material. Mycopathologia, **112**: 81-92.

Voss KA, Chamberlain WJ, Bacon CW, & Norred WP (1993) A preliminary investigation on renal and hepatic toxicity in rats fed purified fumonisin B₁. Nat Toxins, **1**: 222-228.

Voss KA, Chamberlain WJ, Bacon CW, Herbert RA, Walters DB, & Norred WP (1995) Subchronic feeding study of the mycotoxin fumonisin B₁ in B6C3F₁ mice and Fischer 344 rats. Fundam Appl Toxicol, **24**: 102-110.

Voss KA, Bacon CW, Norred WP, Chapin RE, Chamberlain WJ, Plattner RD, & Meredith FI (1996a) Studies on the reproductive effects of *Fusarium moniliforme* culture material in rats and the biodistribution of [14C]fumonisin B$_1$ in pregnant rats. Nat Toxins, **4**: 24-33.

Voss KA, Riley RT, Bacon CW, Chamberlain WJ, & Norred WP (1996b) Subchronic toxic effects of *Fusarium moniliforme* and fumonisin B$_1$ in rats and mice. Nat Toxins, **4**: 16-23.

Voss KA, Bacon CW, Meredith FI, & Norred WP (1996c) Comparative subchronic toxicity studies of nixtamalized and water-extracted *Fusarium moniliforme* culture material. Food Chem Toxicol, **34**: 623-632.

Voss KA, Riley RT, Bacon CW, Meredith FI, & Norred WP (1998) Toxicity and sphinganine levels are correlated in rats fed fumonisin B$_1$ (FB$_1$) or hydrolyzed FB . Environ Toxicol Pharmacol, **5**: 101-104.

Vudathala DK, Prelusky DB, Ayroud M, Trenholm HL, & Miller JD (1994) Pharmacokinetic fate and pathological effects of ^{14}C-fumonisin B$_1$ in laying hens. Nat Toxins, **2**: 81-88.

Wang D-S, Sugiura Y, Ueno Y, Buddhanont P, & Suttajit M (1993) A limited survey for the natural occurrence of fumonisins in Thailand. Thai J Toxicol, **9**: 42-46.

Wang E, Norred WP, Bacon CW, Riley RT, & Merrill AH Jr (1991) Inhibition of sphingolipid biosynthesis by fumonisins: implications for diseases associated with *Fusarium moniliforme*. J Biol Chem, **266**: 14486-14490.

Wang E, Ross PF, Wilson TM, Riley RT, & Merrill AH Jr (1992) Increases in serum sphingosine and sphinganine and decreases in complex sphingolipids in ponies given feed containing fumonisins, mycotoxins produced by *Fusarium moniliforme*. J Nutr, **122**: 1706-1716.

Wang D-S, Liang Y-X, Chau NT, Dien LD, Tanaka T, & Ueno Y (1995) Natural co-occurrence of *Fusarium* toxins and aflatoxin B$_1$ in corn for feed in North Vietnam. Nat Toxins, **3**: 445-449.

Wang H, Jones C, Ciacci-Zanella J, Holt T, Gilchrist DG, & Dickman MB (1996) Fumonisins and *Alternaria alternata lycopersici* toxins: Sphinganine analog mycotoxins induce apoptosis in monkey kidney cells. Proc Natl Acad Sci (USA), **93**: 3461-3465.

Wang E, Riley RT, Meredith FI, & Merrill AH Jr (1999) Fumonisin B$_1$ consumption by rats causes reversible, dose-dependent increases in urinary sphinganine and sphingosine. J Nutr, **129**: 214-220.

Ware GM, Francis O, Kuan SS, Umrigar P, Carman A, Carter L, & Bennett GA (1993) Determination of fumonisin B$_1$ in corn by high performance liquid chromatography with fluorescence detection. Anal Lett, **26**: 1751-1770.

Ware GM, Umrigar PP, Carman AS Jr, & Kuan SS (1994) Evaluation of fumonitest immunoaffinity columns. Anal Lett, **27**: 693-715.

Wattenberg EV, Badria FA, & Shier WT (1996) Activation of mitogen-activated protein kinase by the carcinogenic mycotoxin fumonisin B$_1$. Biochem Biophys Res Commun, **227**: 622-627.

Weibking TS, Ledoux DR, Brown TP, & Rottinghaus GE (1993a) Fumonisin toxicity in turkey poults. J Vet Diagn Invest, **5**: 75-83.

Weibking TS, Ledoux DR, Bermudez AJ, Turk JR, Rottinghaus GE, Wang E, & Merrill AH Jr (1993b) Effects of feeding *Fusarium moniliforme* culture material, containing known levels of fumonisin B₁, on the young broiler chick. Poult Sci, **72**: 456-466.

Weibking T, Ledoux DR, Bermudez AJ, Turk JR, & Rottinghaus GE (1995) Effects on turkey poults of feeding *Fusarium moniliforme* M-1325 culture material grown under different environmental conditions. Avian Dis, **39**: 32-38.

Wieder T, Orfanos CE, & Geilen CC (1998) Induction of ceramide-mediated apoptosis by the anticancer phospholipid analog, hexadecylphosphocholine. J Biol Chem, **273**: 11025-11031.

Wild CP, Castegnaro M, Ohgaki H, Garren L, Galendo D, & Miller JD (1997) Absence of a synergistic effect between fumonisin B₁ and N-nitrosomethylbenzylamine in the induction of oesophageal papillomas in the rat. Nat Toxins, **5**: 126-131.

Wilson BJ & Maronpot RR (1971) Causative fungus agent of leukoencephalomalacia in equine animals. Vet Rec, **88**: 484-486.

Wilson TM, Nelson PE, & Knepp CR (1985) Hepatic neoplastic nodules, adenofibrosis, and cholangiocarcinomas in male Fisher 344 rats fed corn naturally contaminated with *Fusarium moniliforme*. Carcinogenesis, **6**: 1155-1160.

Wilson TM, Ross PF, Rice LG, Osweiler GD, Nelson HA, Owens DL, Plattner RD, Reggiardo C, Noon TH, & Pickrell JW (1990) Fumonisin B₁ levels associated with an epizootic of equine leukoencephalomalacia. J Vet Diagn Invest, **2**: 213-216.

Wilson TM, Ross PF, Owens DL, Rice LG, Green SA, Jenkins SJ, & Nelson HA (1992) Experimental reproduction of ELEM–A study to determine the minimum toxic dose in ponies. Mycopathologia, **117**: 115-120.

Witty JP, Bridgham JT, & Johnson AL (1996) Induction of apoptotic cell death in hen granulosa cells by ceramide. Endocrinology, **137**: 5269-5277.

Wu W-I, McDonough VM, Nickels JT, Ko SJ, Fischl AS, Vales TR, Merrill AH Jr, & Carman GM (1995) Regulation of lipid biosynthesis in *Saccharomyces cerevisiae* by fumonisin B₁. J Biol Chem, **270**: 13171-13178.

Xu J, Yeh C-H, Chen S, He L, Sensi SL, Canzoniero LMT, Choi DW, & Hsu CY (1998) Involvement of *de novo* ceramide biosynthesis in tumor necrosis factor-α/cycloheximide-induced cerebral endothelial cell death. J Biol Chem, **273**: 16521-16526.

Yamashita A, Yoshizawa T, Aiura Y, Sanchez P, Dizon EI, Arim RH, & Sardjono (1995) *Fusarium* mycotoxins (fumonisins, nivalenol and zearalenone) and aflatoxins in corn from southeast Asia. Biosci Biotechnol Biochem, **59**: 1804-1807.

Yeung JM, Wang H-Y, & Prelusky DB (1996) Fumonisin B₁ induces protein kinase C translocation via direct interaction with diacylglycerol binding site. Toxicol Appl Pharmacol, **141**: 178-184.

Yin J-J, Smith MJ, Eppley RM, Troy AL, Page SW, & Sphon JA (1996) Effects of fumonisin B_1 and (hydrolyzed) fumonisin backbone AP_1 on membranes: A spin-label study. Arch Biochem Biophys, **335**: 13-22.

Yin J-J, Smith MJ, Eppley RM, Page SW, & Sphon JA (1998) Effects of fumonisin B1 on lipid peroxidation in membranes. Biochim Biophys Acta, **1371**: 134-142.

Yoo H-S, Norred WP, Wang E, Merrill AH Jr, & Riley RT (1992) Fumonisin inhibition of *de novo* sphingolipid biosynthesis and cytotoxicity are correlated in LLC-PK₁ cells. Toxicol Appl Pharmacol, **114**: 9-15.

Yoo H-S, Norred WP, Showker JL, & Riley RT (1996) Elevated sphingoid bases and complex sphingolipid depletion as contributing factors in fumonisin-induced cytotoxicity. Toxicol Appl Pharmacol, **138**: 211-218.

Yoshizawa T, Yamashita A, & Luo Y (1994) Fumonisin occurrence in corn from high- and low-risk areas for human esophageal cancer in China. Appl Environ Microbiol, **60**: 1626-1629.

Young JC & Lafontaine P (1993) Detection and characterization of fumonisin mycotoxins as their methyl esters by liquid chromatography/particle-beam mass spectrometry. Rapid Commun Mass Spectrom, **7**: 352-359.

Zacharias C, Van Echten-Deckert G, Wang E, Merrill AH Jr, & Sandoff K (1996) The effect of fumonisin B_1 on developing chick embryos: Correlation between *de novo* sphingolipid biosynthesis and gross morphological changes. Glycoconj J, **13**: 167-175.

Zhang H, Buckley NE, Gibson K, & Spiegel S (1990) Sphingosine stimulates cellular proliferation via a protein kinase C-independent pathway. J Biol Chem, **265**: 76-81.

Zhang H, Desai NN, Olivera A, Seki T, Booker G, & Spiegel S (1991) Sphingosine-1-phosphate, a novel lipid, involved in cellular proliferation. J Cell Biol, **114**: 155-167.

Zhang H, Nagashima H, & Goto T (1997) Natural occurrence of mycotoxins in corn, samples from high and low risk areas for human esophageal cancer in China. Mycotoxins, **44**: 29-35.

Zhen YZ (1984) [The culture and isolation of fungi from the cereals in five high and three low incidence counties of esophageal cancer in Henan Province.] J Chin Tumor, **6**: 27-29 (in Chinese).

Zoller O, Sager F, & Zimmerli B (1994) [Occurrence of fumonisins in food.] Mitt Gebiete Lebensm Hyg, **85**: 81-99 (in German).

APPENDIX 1. NATIONAL GUIDELINES FOR FUMONISINS

The US Food and Drug Administration, Center for Veterinary Medicine unofficial guidelines (Miller et al., 1996) recommend that the non-roughage portion of feeds for equine species should be less than 5 ppm FB_1 (< 5 mg/kg), for porcine species the total diet should contain less than 10 ppm FB_1 (< 10 mg/kg), for beef cattle the non-roughage portion should be less than 50 ppm FB_1 (< 50 mg/kg), and for poultry the complete feed should contain less than 50 ppm FB_1 (< 50 mg/kg).

An official tolerance value for dry maize products (1 mg/kg FB_1 plus FB_2) has been issued in Switzerland (Swiss Federal Office of Public Health, 1997).

APPENDIX 2. NATURAL OCCURRENCE OF FUMONISIN B₁ (FB₁) IN MAIZE-BASED PRODUCTS

Product	Country	Positive/total	FB₁ (mg/kg)	Reference
North America				
maize	Canada	1/1	0.08	Stack & Eppley (1992)
maize	Canada	9/98	<1–2.5	Miller et al. (1995)
maize flour	Canada	1/2	0.05	Sydenham et al. (1991)
maize feed	USA	3/3	37–122	Wilson et al. (1990)
maize feed	USA	2/2	12–130	Plattner et al. (1990)
maize feed	USA	81/93	<1–126	Ross et al. (1991a)
maize feed	USA	158/213	<1–330	Ross et al. (1991b)
maize feed	USA	15/15	1.3–5.2	Thiel et al. (1991b)
maize feed	USA	2/2	105–155	Colvin & Harrison (1992)
maize feed	USA	29/29	3–330	Osweiler et al. (1992)
maize feed	USA	20/21	<1–73 [a]	Bane et al. (1992)
maize screenings (feed)	USA	6/6	1.7–196.5	Stack & Eppley (1992)
maize feed	USA	1/1	86.0	Park et al. (1992)
maize feed	USA	14/14	1.3–27.0	Sydenham et al. (1992)
maize feed	USA	160/160	0.1–239	Murphy et al. (1993)
maize feed	USA	85/85	2.6–32 [a]	Price et al. (1993)
maize feed	USA	5/5	0.22–1.41	Hopmans & Murphy (1993)
maize feed	USA	0/29	–	Chamberlain et al. (1993)
maize feed	USA	0/1	–	Holcomb et al. (1993)
maize feed	USA	5/5	0.77–6.2	Rumbeiha & Oehme (1997)
maize	USA	6/6	0.14–16.31	Stack & Eppley (1992)
maize	USA	13/99	1.2–3.2	Price et al. (1993)
maize	USA	155/175	< 1–37.9	Murphy et al. (1993)
maize	USA	24/28	av. 0.87	Chamberlain et al. (1993)

Appendix 2 (contd),

Product	Country	Positive/ total	FB$_1$ (mg/kg)	Reference
maize	USA	116/322	1–> 10	Shelby et al. (1994a)
maize flour	USA	13/25	0.05–0.35	Rumbeiha & Oehme (1997)
maize flour	USA	7/7	0.40–6.32	Pestka et al. (1994)
maize flour	USA	15/16	0.05–2.79	Sydenham et al. (1991)
maize flour	USA	16/16	0.28–2.05	Stack & Eppley (1992)
maize flour	USA	6/6	0.21–0.84	Hopmans & Murphy (1993)
maize grits	USA	10/10	0.11–2.55	Sydenham et al. (1991)
maize grits	USA	5/5	0.14–0.27	Stack & Eppley (1992)
maize flakes	USA	0/2	–	Sydenham et al. (1991)
maize flakes	USA	2/5	0.01	Stack & Eppley (1992)
maize cereals (bran, fibre, pops)	USA	7/12	0.06–0.33	Stack & Eppley (1992)
popcorn	USA	1/18	0.07	Rumbeiha & Oehme (1997)
sweet maize	USA	1/1	0.026	Hopmans & Murphy (1993)
sweet maize	USA	37/97	0.004–0.35	Trucksess et al. (1995)
tortillas	USA	1/3	0.05–0.06	Sydenham et al. (1991)
miscellaneous maize foods [b]	USA	4/4	0.09–0.70	Sydenham et al. (1991)
miscellaneous maize foods [b]	USA	6/11	0.01–0.12	Stack & Eppley (1992)
miscellaneous maize foods [b]	USA	4/4	0.02–0.32	Hopmans & Murphy (1993)
miscellaneous maize foods [b]	USA	3/5	0.05–1.21	Pestka et al. (1994)

Appendix 2 (contd),

Product	Country	Positive/ total	FB$_1$ (mg/kg)	Reference
Latin America				
maize	Argentina	17/17	1.11–6.70	Sydenham et al. (1993b)
maize	Argentina	51/51	0.18–27.05	Visconti et al. (1995); Ramirez et al. (1996)
maize feed	Brazil	20/21	0.2–38.5	Sydenham et al. (1992)
maize	Brazil	47/48	0.6–18.5	Hirooka et al. (1996)
tortillas	Texas-Mexico border	50/52	av. 0.19	Stack (1998)
masas	Texas-Mexico border	8/8	av. 0.26	Stack (1998)
maize flour	Peru	1/2	0.66	Sydenham et al. (1991)
alkali-treated kernels	Peru	0/2	–	Sydenham et al. (1991)
maize feed	Uruguay	13/13	0.26–6.3	Piñeiro et al. (1997)
maize	Uruguay	11/22	0.17–3.7	Piñeiro et al. (1997)
maize snacks	Uruguay	4/10	0.15–0.31	Piñeiro et al. (1997)
frozen maize	Uruguay	1/7	0.16	Piñeiro et al. (1997)
polenta	Uruguay	3/12	0.1–0.43	Piñeiro et al. (1997)
maize flour	Venezuela	1/1	0.07	Stack & Eppley (1992)
Europe				
maize	Austria	6/9	1–15	Lew et al. (1991)
maize flour	Austria	–/3[f]	0.05–1.15	Sydenham et al. (1993a)
maize flour	Bulgaria	–/15[f]	0.05–0.21	Sydenham et al. (1993a)
maize	Croatia	11/19	0.01–0.06	Doko et al. (1995)
maize flour	Czech	22/22	0.01–0.49[a]	Ostrý & Ruprich (1998)
maize pastes	Czech	6/11	0.01–0.51[a]	Ostrý & Ruprich (1998)
maize-extruded bread	Czech	30/35	0.01–1.8[a]	Ostrý & Ruprich (1998)

Appendix 2 (contd),

Product	Country	Positive/ total	FB$_1$ (mg/kg)	Reference
polenta	Czech	6/7	0.01–1.2 [a]	Ostrý & Ruprich (1998)
porridge	Czech	18/19	0.01–0.79 [a]	Ostrý & Ruprich (1998)
maize feed	France	43/58	0.02–8.82	Doko et al. (1994)
maize feed	France	35/35	0.02–2.17	Dragoni et al. (1996)
maize flour	France	1/1	1.24	Sydenham et al. (1993a)
miscellaneous maize foods [b]	France	10/22	0.02–1.50	Visconti et al. (1995)
maize	Germany	49/458	0.007–4.83	Meister et al. (1996)
imported maize	Germany	21/21	0.014–1.11 [a]	Meister et al. (1996)
maize grits	Germany	1/2	0.01	Usleber et al. (1994a)
grits flour, semolina	Germany	60/71	0.01–16.00	Usleber et al. (1994b) [c]
semolina	Germany	10/11	0.01–1.23	Usleber et al. (1994a)
popcorn	Germany	13/29	0.01–0.16	Usleber et al. (1994b) [c]
popcorn	Germany	4/6	0.01–0.11	Usleber et al. (1994a)
infant foods	Germany	0/91	–	Usleber et al. (1994b) [c]
maize	Hungary	56/92	0.05–75.10	Fazekas et al. (1998)
maize feed	Italy	23/25	0.02–8.40	Minervini et al. (1992)
maize screen (feed)	Italy	3/3	55.2–70.0	Caramelli et al. (1993)
maize feed	Italy	1/1	60	Doko & Visconti (1994)
maize	Italy	7/7	0.1–5.3	Doko & Visconti (1994)
maize	Italy	6/6	125–250	Bottalico et al. (1995)
maize genotypes	Italy	26/26	0.01–2.33	Doko & Visconti (1994)
commercial maize kernels	Italy	7/7	0.10–5.31	Doko & Visconti (1994)
maize flour	Italy	7/7	0.42–3.73	Doko & Visconti (1994)
maize grits	Italy	1/1	3.76	Doko & Visconti (1994)
polenta	Italy	20/20	0.15–3.76	Pascale et al. (1995)

Appendix 2 (contd),

Product	Country	Positive/ total	FB$_1$ (mg/kg)	Reference
popcorn, maize flakes, tortilla chips	Italy	6/10	0.01–0.06	Doko & Visconti (1994)
extruded maize	Italy	6/6	0.79–6.10	Doko & Visconti (1994)
sweet maize	Italy	5/5	0.06–0.79	Doko & Visconti (1994)
maize flour	Netherlands	5/7	0.008–0.09	de Nijs et al. (1998c)
maize for bread	Netherlands	8/19	0.008–0.38	de Nijs et al. (1998c)
maize for popcorn	Netherlands	2/10	0.008–0.11	de Nijs et al. (1998c)
maize foods	Netherlands	12/42	0.008–1.43	de Nijs et al. (1998c)
imported maize	Netherlands	61/62	0.03–3.35	de Nijs et al. (1998b)
maize	Poland	2/7	0.01–0.02	Doko et al. (1995)
maize	Portugal	9/9	0.09–2.30	Doko et al. (1995)
maize	Romania	3/6	0.01–0.02	Doko et al. (1995)
maize feed	Spain	136/171	av. 3.3	Castella et al. (1997)
maize kernels	Spain	1/1	0.72	Visconti et al. (1995)
maize flour	Spain	1/3	0.05–0.07	Sanchis et al. (1994)
maize flour	Spain	16/17	≤ 0.50[a]	Burdaspal & Legarda (1996)
maize grits	Spain	3/15	0.05–0.09	Sanchis et al. (1994)
maize flakes	Spain	2/12	0.05–0.10	Sanchis et al. (1994)
maize starch	Spain	1/13	0.03	Burdaspal & Legarda (1996)
sweet maize	Spain	3/3	0.72	Visconti et al. (1995)
miscellaneous maize foods[b]	Spain	2/20	0.05–0.20	Sanchis et al. (1994)
maize flour	Sweden	1/1	0.13	Visconti et al. (1995)
popcorn	Sweden	1/1	0.13	Visconti et al. (1995)
maize feed	Switzerland	6/22	av. 0.24	Pittet et al. (1992)
maize flour	Switzerland	2/7	av. 0.09	Pittet et al. (1992)
maize grits[d]	Switzerland	34/55	av. 0.26	Pittet et al. (1992)
maize grits, flour[d]	Switzerland	27/27	0.01–2.20	Zoller et al. (1994)
maize flakes	Switzerland	1/12	0.06	Pittet et al. (1992)
popcorn	Switzerland	8/13	0.005–0.25	Zoller et al. (1994)

Appendix 2 (contd),

Product	Country	Positive/ total	FB$_1$ (mg/kg)	Reference
sweet maize	Switzerland	1/7	0.07	Pittet et al. (1992)
miscellaneous maize foods[b]	Switzerland	0/17	–	Pittet et al. (1992)
maize feed	UK	24/29	0.05–4.55	Scudamore & Chan (1993)
maize	UK	65/67	0.03–24	Scudamore et al. (1998)
polenta	UK	16/20	0.02–2.12[a]	Patel et al. (1997)
maize snacks	UK	31/40	0.01–0.22[a]	Patel et al. (1997)
popcorn	UK	6/22	0.01–0.78[a]	Patel et al. (1997)
Africa				
maize	Benin	9/11	0.02–2.63	Doko et al. (1995)
maize flour	Botswana	5/5	0.18–0.45	Sydenham et al. (1993a)
miscellaneous maize foods[b]	Botswana	6/6	0.03–0.35	Doko et al. (1996)
maize flour	Egypt	2/2	1.78–2.98	Sydenham et al. (1991)
maize kernels	Kenya	1/1	0.78	Doko et al. (1996)
maize flour	Kenya	–/3[f]	0.05–0.11	Sydenham et al. (1993a)
maize kernels	Malawi	7/8	0.02–0.11	Doko et al. (1996)
maize kernels	Mozambique	3/3	0.24–0.29	Doko et al. (1996)
maize feed	South Africa	15/15	0.47–4.34	Viljoen et al. (1994)
mixed feed	South Africa	1/1	8.85	Shephard et al. (1990)
maize	South Africa	3/3	10–83	Sydenham et al. (1990a)
maize	South Africa	60/60	0.2–46.9	Sydenham et al. (1990b)
maize	South Africa[e]	62/74	0.05–117.5	Rheeder et al. (1992)
maize	South Africa	24/68	0.05–0.87	Sydenham (1994)[c]
maize flour	South Africa	46/52	0.05–0.48	Sydenham et al. (1991)
maize flour	South Africa	2/2	0.06–0.07	Doko et al. (1996)
maize grits	South Africa	10/18	0.05–0.19	Sydenham et al. (1991)
maize flakes	South Africa	0/3	–	Sydenham et al. (1991)

Appendix 2 (contd),

Product	Country	Positive/ total	FB$_1$ (mg/kg)	Reference
miscellaneous maize foods [b]	South Africa	2/8	0.05–0.09	Sydenham et al. (1991)
maize kernels	Tanzania	8/9	0.02–0.16	Doko et al. (1996)
maize kernels	Uganda	1/1	0.60	Doko et al. (1996)
maize	Zambia	20/20	0.02–1.42	Doko et al. (1995)
maize flour	Zambia	1/1	0.74	Doko et al. (1996)
maize kernels	Zimbabwe	1/2	0.12	Doko et al. (1996)
maize flour	Zimbabwe	3/3	1.06–3.63	Sydenham et al. (1993a)
maize flour	Zimbabwe	4/4	0.05–1.91	Doko et al. (1996)
Asia				
maize	China	2/5	5.3–8.4	Ueno et al. (1993)
maize	China	18/47	0.18–2.9	Yoshizawa et al. (1994)
maize	China	27/68	0.01–1.4	Kang et al. (1994)
maize	China [e]	34/34	18–155	Chu & Li (1994)
maize	China [e]	134/240	0.08–34.87	Ueno et al. (1997)
maize	China [e]	37/54	0.08–21	Gao & Yoshizawa (1997)
maize flour	China	0/3	–	Sydenham et al. (1993a)
maize flour	China	3/4	0.06–0.2	Ueno et al. (1993)
gluten	India	1/1	0.7	Scudamore & Chan (1993)
maize	Indonesia	7/12	0.05–1.8	Yamashita et al. (1995)
maize	Indonesia	16/16	0.05–2.44	Ali et al. (1998)
maize grits	Japan	14/17	0.20–2.60	Ueno et al. (1993)
sweet maize	Japan	0/8	–	Ueno et al. (1993)
maize snack	Japan	0/31	–	Ueno et al. (1993)
maize soup	Japan	0/7	–	Ueno et al. (1993)
maize feed	Korea	5/12	0.05–1.33	Lee et al. (1994)
maize	Nepal	12/24	0.05–4.6	Ueno et al. (1993)
maize	Philippines	26/50	0.05–1.8	Yamashita et al. (1995)
maize	Philippines	9/10	0.3–10.0	Bryden et al. (1996)
miscellaneous maize foods [b]	Taiwan	52/153	0.07–2.39	Tseng & Liu (1997)

Appendix 2 (contd),

Product	Country	Positive/ total	FB$_1$ (mg/kg)	Reference
maize	Thailand	19/27	0.06–18.8	Yamashita et al. (1995)
maize feed	Thailand	5/22	0.05–1.59	Wang et al. (1993)
maize flour	Thailand	6/6	0.48–0.88	Wang et al. (1993)
maize grits	Thailand	5/5	0.25–1.82	Wang et al. (1993)
maize	Vietnam	12/12	0.3–9.1	Bryden et al. (1996)
maize	Vietnam	8/15	0.27–3.45	Wang et al. (1995)
maize powder	Vietnam	15/17	0.27–1.52	Wang et al. (1995)
Oceania				
maize	Australia	67/70	0.3–40.6	Bryden et al. (1996)
maize flour	New Zealand	0/12	–	Sydenham et al. (1993a)

[a] Fumonisins B$_1$ + B$_2$ + B$_3$
[b] Includes maize snacks, canned maize, frozen maize, extruded maize, bread, maize-extruded bread, biscuit, cereals, chips, flakes, pastes, starch, sweet maize, infant foods, gruel, purée, noodles popcorn, porridge, tortillas, tortilla chips, masas, popped maize, soup, taco, tostada
[c] From: Shephard et al. (1996a)
[d] Maize grits and flour samples analysed were imported cereals (mainly from Argentina)
[e] Fumonisin levels in low- and high-risk area for human oesophageal cancer
[f] The number of positive samples was not indicated in the original report

RESUME, EVALUATION ET RECOMMANDATIONS

1. Résumé

1.1 Identité, propriétés physique et chimiques et méthodes d'analyse

La fumonisine B_1 (FB_1), de formule brute $C_{34}H_{59}NO_{15}$, est le diester de l'acide propane-1,2,3-tricarboxylique et du 2-amino-12,16-diméthyl-3,5,10,14,15-pentahydroxyeicosane (masse moléculaire relative: 721). C'est la plus abondante des fumonisines, qui constituent une famille de toxines dont on a identifié au moins 15 membres. A l'état pur, ce composé se présente sous la forme d'une poudre hygroscopique de couleur blanche, soluble dans l'eau, le mélange eau-acétonitrile et le méthanol. Elle est stable dans le mélange eau-acétonitrile à 1:1 et instable dans le méthanol. Elle est stable aux températures utilisées pour la transformation des denrées alimentaires ainsi qu'à la lumière.

Plusieurs méthodes d'analyse ont été proposées, notamment la chromatographie sur couche mince ou en phase liquide, la spectrométrie de masse, la chromatographie en phase gazeuse après hydrolyse et un certain nombre de méthodes immunochimiques. En fait, la plupart des dosages se font par chromatographie en phase liquide d'un dérivé fluorescent.

1.2 Sources d'exposition humaine

La FB_1 est produite par plusieurs espèces de *Fusarium*, mais essentiellement par *Fusarium verticillioides* (Sacc.) Nirenberg (= *Fusarium moniloforme* Sheldon) qui est l'un des parasites cryptogamiques les plus fréquents du maïs. La FB_1 peut s'accumuler en quantités importantes dans le maïs lorsque les conditions météorologiques sont favorables à l'apparition de la fusariose.

1.3 Transport, distribution et transformation dans l'environnement

On est fondé à penser que les fumonisines peuvent être métabolisées par certains microorganismes terricoles. On sait toutefois peu de choses de leur devenir dans l'environnement une fois qu'elles ont été excrétées ou que les produits qui en contiennent ont été transformés.

1.4 Concentrations dans l'environnement et exposition humaine

On a mis en évidence de la FB_1 dans le maïs et les produits qui en dérivent partout dans le monde à des concentrations de l'ordre du mg/kg, parfois en association avec d'autres mycotoxines. Des concentrations de cet ordre ont également été observées dans des denrées alimentaires destinées à la consommation humaine. Lors de la mouture à sec du maïs, la fumonisine se répartit dans le son, le germe et la farine. Lors d'essais de mouture par voie humide, on a mis en évidence la toxine dans l'eau de macération, dans le gluten, dans les fibres et les germes, à l'exclusion de l'amidon. La FB_1 reste stable dans le maïs et la polenta, mais elle s'hydrolyse dans les aliments à base de maïs traités par des solutions alcalines à chaud.

La FB_1 est absente du lait, de la viande et des oeufs provenant d'animaux nourris avec du maïs dont la teneur en toxine ne représente aucun danger pour eux. On estime que l'exposition humaine journalière aux Etats-Unis, au Canada, en Suisse, aux Pays-Bas et au Transkei (Afrique du Sud) varie entre 0,017 et 440 µg/kg de poids corporel. On ne possède aucune donnée sur l'exposition professionnelle par inhalation.

1.5 Cinétique et métabolisme chez l'animal

On ne dispose d'aucune donnée sur la cinétique ou le métabolisme de la FB_1 chez l'Homme. Chez les animaux de laboratoire, le composé est peu résorbé après ingestion; il disparaît rapidement du courant sanguin et se retrouve inchangé dans les matières fécales. Il est excrété en quantité importante par la voie biliaire et en faible proportion dans les urines. Chez les primates non humains et certains ruminants, la fumonisine peut subir une dégradation hydrolytique partielle dans l'intestin. Elle subsiste en petite quantité dans le foie et les reins.

1.6   Effets sur l'animal et les systèmes d'épreuve *in vitro*

La FB_1 s'est révélée hépatotoxique pour les espèces animales sur lesquelles elle a été testée, notamment le rat, la souris, les équidés, le lapin, le porc et les primates non humains. Sauf dans le cas du hamster doré, on n'observe d'effets embryotoxiques ou tératogènes que concurremment ou postérieurement aux manifestations toxiques qui se produisent chez la mère. Les fumonisines sont néphrotoxiques chez le porc, le rat, le mouton, la souris et le lapin. Dans le cas du rat et du lapin, la néphrotoxicité se manifeste à des doses plus faibles que l'hépatotoxicité. On sait que la leucoencéphalomalacie équine et l'oedème pulmonaire porcin observés après consommation de provendes à base de maïs sont dus à la présence de fumonisines. Les données dont on dispose sur les propriétés immunologiques de la FB_1 sont limitées. A la dose de 50 mg/kg de nourriture, le composé a provoqué des cancers du foie chez une souche de rats et des cancers du rein chez une autre souche; dans les mêmes conditions de dosage et d'administration, il a également provoqué des cancers du foie chez des souris femelles. Il semble qu'il existe une corrélation entre les effets toxiques sur les organes et l'apparition de cancers. La FB_1 a été le premier inhibiteur du métabolisme des sphingolipides qui ait été découvert et on l'utilise beaucoup depuis lors pour étudier le rôle des sphingolipides dans la régulation cellulaire. La FB_1 inhibe la croissance cellulaire; elle entraîne l'accumulation de bases sphingoïdes libres et modifie le métabolisme des lipides chez les animaux, les plantes et certaines levures. Elle ne provoque pas de mutation génique chez les bactéries et mise en présence d'une culture primaire d'hépatocytes de rat, elle n'entraîne pas une synthèse non programmée de l'ADN. On a constaté par contre qu'elle pouvait provoquer des aberrations chromosomiques à faible dose dans ces mêmes cultures cellulaires.

1.7   Effets sur l'Homme

On ne dispose d'aucune donnée confirmée relative à la toxicité aiguë des fumosinines pour l'Homme. Des études effectuées au Transkei (Afrique du Sud) sur la corrélation entre divers effets toxiques et la présence de fumonisines dans la ration alimentaire suggèrent l'existence d'un lien entre ces composés et le cancer de l'oesophage. Cette corrélation a été observée dans des circonstances où il y avait une forte   exposition   aux   fumonisines   et   où   les   conditions environnementales étaient favorables à une importante accumulation de toxines dans le maïs, qui constitue une denrée alimentaire de base dans la région. Des études du même genre ont également été effectuées en Chine. Ces dernières n'ont toutefois pas permis de dégager une relation claire entre la contamination par les fumonisines ou *F. verticillioides* et le cancer de l'oesophage. Fautes de données concernant l'exposition aux fumonisines, il n'est pas possible non plus de tirer de conclusions d'une étude cas-témoins effectuée en Italie sur des sujets de sexe masculin et qui révèle l'existence d'une association entre la consommation de maïs et les cancers des voies digestives supérieures chez les gros buveurs.

Il n'existe pas de marqueurs biologiques valables de l'exposition à la FB_1.

1.8 Effets sur les autres organismes en laboratoire

La FB_1 inhibe la croissance cellulaire et provoque l'accumulation de bases sphingoïdes libres ainsi que la modification du métabolisme des lipides chez *Saccharomyces cerevisiae*.

La FB_1 est phytotoxique, elle provoque des lésions de la membrane cellulaire et réduit la synthèse chlorophyllienne. Elle perturbe également la biosynthèse des sphingolipides chez les végétaux et pourrait jouer un rôle dans les maladies du maïs dues aux espèces de *Fusarium* qui produisent des fumonisines

2. Evaluation des risques pour la santé humaine

2.1 Exposition

L'Homme est exposé partout dans le monde puisque la FB_1 est présente dans le maïs destiné à la consommation humaine. Toutefois, des différences notables existent entre les régions du culture de cette céréale. Cette constatation s'impose lorsque l'on compare pays développés et pays en développement. Par exemple, même si aux Etats-Unis, au Canada et en Europe occidentale la FB_1 est présente dans les produits tirés du maïs, la consommation de ces produits y reste à un niveau modeste. Dans certaines régions d'Afrique, d'Amérique centrale et d'Asie, il y a des populations dont l'apport calorique alimentaire est constitué pour une large part de farine de maïs et la contamination par la FB_1 peut y être importante (voir Appendice 2). Le maïs naturellement contaminé par la FB_1 peut également l'être par d'autres toxines de *F. verticillioides* ou *F. proliferatum* ou encore par des toxines d'importance agricole telles que le désoxynivalénol, la zéaralénone, l'aflatoxine ou l'ochratoxine.

Les procédés de transformation des denrées alimentaires utilisés en Amérique du Nord ou en Europe occidentale n'ont aucun effet sur la stabilité de la FB_1. Le traitement du maïs par une base ou son lavage à l'eau réduisent sensiblement la teneur en toxine. Cependant on constate toujours des effets hépatotoxiques et néphrotoxiques chez les animaux de laboratoire. On sait peu de choses sur la manière dont les techniques de préparation des aliments utilisées dans le monde en développement peuvent agir sur la FB_1 présente dans les produits tirés du maïs.

2.2 Nature des dangers

Le rôle étiologique de la FB_1 dans la leucoencéphalomalacie équine est établi. D'importantes flambées de cette zoonose mortelle se sont produites aux Etats-Unis au cours du 19ème siècle et plus récemment en 1989-1990. De même, on a également montré que cette toxine était à l'origine d'une zoonose tout aussi mortelle, l'oedème pulmonaire porcin. Une faible dose de FB_1 peut également être mortelle pour le lapin, comme on a pu l'observer sur des lapines gravides. Chez toutes les espèces animales étudiées, y compris les primates non humains, on a constaté que cette toxine provoquait des lésions rénales et hépatiques. La FB_1 provoque une hyper-cholestérolémie chez plusieurs espèces animales, dont les primates non humains. On de bonnes raisons de penser que dans les maladies animales dues à une exposition à la FB_1, il y a modification du métabolisme des lipides. Les effets toxiques observés *in vivo* ou *in vitro* sont précédés ou accompagnés d'une perturbation du métabolisme des sphingolipides. L'utilisation des fumonisines dans l'étude de la fonction des sphingolipides a montré que ces composés sont nécessaires à la croissance cellulaire et qu'il peuvent affecter les molécules signal de différentes manières, avec pour conséquence la mort cellulaire par apoptose ou nécrose, la différenciation cellulaire ou encore la modification de la réponse immunitaire. Après exposition à la FB₁, on observe communément une modification du métabolisme des lipides et de l'expression ou de l'activité des enzymes qui jouent un rôle clé dans la progression du cycle cellulaire. La FB₁ n'exerce pas d'effets toxiques sur le développement chez le rat, la souris ou le lapin. En revanche, elle est foetotoxique chez le hamster doré à forte dose même quand elle n'a pas d'effet sur la mère.

Chez les rongeurs, la cancérogénicité de la FB₁ varie selon l'espèce, la souche et le sexe. La seule étude qui ait été effectuée sur des souris B6C3F₁ a montré que la toxine provoquait des cancers du foie chez les femelles à la dose de 50 mg/kg de nourriture. Des cancers primitifs du foie et des cholangiomes ont été observés chez des rats mâles BD IX qui avaient reçu, pendant une durée allant jusqu'à 26 mois, une alimentation contenant 50 mg/kg de FB₁. Chez des rats mâles F344/N alimentés dans les mêmes conditions de dosage, on a mis en évidence des adénomes et des carcinomes au niveau des tubules rénaux. Il semble qu'il y ait corrélation entre la toxicité pour tel ou tel organe et l'apparition de cancers à ce niveau.

Les études de génotoxicité sont en nombre limité. Celles qui portent sur des bactéries n'ont pas révélé d'effets mutagènes. Dans des cultures de cellules mammaliennes on n'a pas non plus décelé de synthèse non programmée de l'ADN mais selon une étude, la FB₁ a provoqué des cassures chromosomiques dans des hépatocytes de rat. Selon d'autres travaux, la FB₁ accroît la peroxydation des lipides *in vivo* et *in vitro*. Il est possible qu'il existe une relation de cause à effet entre la peroxydation des lipides et les cassures chromosomiques.

Les teneurs en FB₁ supérieures à 100 mg/kg constatées dans le maïs consommé par l'Homme en Afrique et en Chine pourraient probablement provoquer selon le cas, des leucoencéphalomalacies, des oedèmes pulmonaires et des cancers si on donnait à manger ce maïs à des chevaux, des porcs, des rats ou des souris. On connaît des cas où l'exposition humaine à cette toxine est très importante, mais aucun cas d'intoxication aiguë par une fumonisine n'a été décrit. Les études de corrélation effectuées au Transkei (Afrique du Sud) incitent à penser qu'il pourrait y avoir un lien entre une exposition à la fumonisine par voie alimentaire et le cancer de l'oesophage. On a effectivement constaté un taux élevé de cancers de l'oesophage là où l'exposition à cette toxine était importante et où les conditions environnementales étaient favorables à l'accumulation de fumonisines dans le maïs, qui constitue un aliment de base dans ces régions.

Une étude cas-témoins réalisée en Italie a mis en évidence une association entre la consommation de maïs et les cancers des voies digestives supérieures, et notamment le cancer de l'oesophage chez les gros buveurs. Aucune donnée sur l'exposition à la toxine n'a été fournie.

2.3 Relation dose-réponse

La dose la plus faible capable de provoquer l'apparition de cancers du foie chez l'animal de laboratoire est égale à 50 mg par kg de nourriture dans le cas de rats mâles BD IX et de souris femelles B6C3F1/Nctr; aucun cancer n'a été observé aux doses respectives de 25 ou 15 mg/kg de nourriture. Dans chaque cas et avec les mêmes souches de rats et de souris, on a observé des signes d'hépatotoxicité ou de modification du métabolisme des lipides à ces mêmes doses ou à des doses inférieures. La dose la plus faible ayant provoqué des cancers du rein chez des rats mâles de souche F344/Nctr était égale à 50 mg/ kg de nourriture; aucun effet cancérogène n'a été observé à la dose de 15 mg/kg de nourriture. Ces études ont également montré qu'à des doses inférieures à celles qui provoquaient l'apparition de cancers, il y avait apoptose et prolifération des cellules des tubules rénaux, avec des modifications au niveau des sphingolipides tissulaires et urinaires.

On ne possède aucune donnée qui puisse permettre d'établir une relation quantitative entre l'exposition à la FB$_1$ et d'éventuels effets sur l'organisme humain.

2.4 Caractérisation du risque

La FB$_1$ est cancérogène pour la souris et le rat et elle provoque des maladies mortelles chez le porc et le cheval à des concentrations auxquelles l'Homme pourrait être exposé. Le Groupe de travail n'a pas été en mesure de formuler une estimation quantitative du risque pour la santé humaine mais il a été d'avis qu'il y avait urgence à cet égard.

3. Recommandations pour la protection de la santé humaine

a) Il faudrait établir les limites d'exposition par voie alimentaire. On devra s'attacher en particulier aux populations dont l'apport calorique provient pour une grande part de la farine de maïs.

b) Des mesures devraient être prises pour limiter l'exposition aux fumonisines et la contamination du maïs par ces toxines; elles pourraient consister

- à changer de culture là où le maïs n'est pas vraiment adapté
- à mettre au point des variétés de maïs qui résistent à la fusariose
- à mieux gérer les cultures
- à éliminer les grains parasités

c) On veillera à faire prendre conscience suffisamment tôt du risque de contamination alimentaire en faisant en sorte qu'il y ait une meilleure communication entre les vétérinaires et les responsables de la santé publique en cas de flambées de mycotoxicoses parmi les animaux domestiques.

d) Il faudrait mettre au point une méthode de recherche de la contamination du maïs par les fumonisines qui soit bon marché, simple et peu sensible aux conditions d'application.

RESUMEN, EVALUACION Y
RECOMENDACIONES

1. Resumen

1.1 Identidad, propiedades físicas y químicas y métodos analíticos

La fumonisina B_1 (FB_1) tiene la fórmula empírica $C_{34}H_{59}NO_{15}$ y es el diéster del ácido propano-1,2,3-tricarboxílico y el 2-amino-12, 16-dimetil-3, 5, 10, 14, 15-pentahidroxieicosano (masa molecular relativa: 721). Es la más frecuente de las fumonisinas, familia de toxinas con 15 miembros identificados por lo menos. La sustancia pura es un polvo higroscópico blanco, soluble en agua, acetonitrilo-agua o metanol, estable en acetonitrilo-agua (1:1), inestable en metanol y estable a la temperatura de elaboración de los alimentos y la luz.

Se han notificado varios métodos analíticos, en particular la cromatografía en capa fina y la cromatografía líquida, la espectrometría de masas, la cromatografía de gases poshidrólisis y métodos inmuno-químicos, aunque la mayoría de los estudios se han realizado utilizando análisis por cromatografía líquida de un derivado fluorescente.

1.2 Fuentes de exposición humana

La FB_1 se produce en varias especies de *Fusarium*, principalmente *Fusarium verticillioides* (Sacc.) Niremberg (= *Fusarium moniliforme* Sheldon), que es uno de los hongos más comunes asociados con el maíz en todo el mundo. En el maíz hay una acumulación importante de FB_1 cuando las condiciones climatológicas favorecen la podredumbre del grano debida a *Fusarium*.

1.3 Transporte, distribución y transformación en el medio ambiente

Hay pruebas de que algunos microorganismos del suelo pueden metabolizar las fumonisinas. Sin embargo, se sabe poco acerca del destino de las fumonisinas en el medio ambiente tras la excreción o la transformación.

1.4 Niveles en el medio ambiente y exposición humana

Se ha detectado FB_1 en el maíz y sus productos en todo el mundo en concentraciones de varios mg/kg, a veces combinada con otras micotoxinas. Se han notificado también concentraciones de mg/kg en alimentos para consumo humano. Como consecuencia de la elaboración en seco del maíz, la fumonisina se distribuye en el salvado, los gérmenes y la harina. En la elaboración en húmedo experimental, se detectó fumonisina en el agua de remojo, el glúten, la fibra y los gérmenes, pero no en el almidón. La FB_1 es estable en el maíz y la polenta, mientras que se hidroliza en los alimentos nixtamalizados a base de maíz, es decir, alimentos elaborados con soluciones alcalinas calientes.

La FB_1 no está presente en la leche, la carne o los huevos de animales alimentados con grano que contiene FB_1 en concentraciones que no afectarían a la salud de los animales. Las estimaciones de la exposición humana para los Estados Unidos, el Canadá, Suiza, los Países Bajos y el Transkei (Sudáfrica) oscilaron entre 0,017 y 440 µg/kg de peso corporal al día. No se dispone de datos sobre la exposición por inhalación en el puesto de trabajo.

1.5 Cinética y metabolismo en los animales

No hay informes sobre la cinética o el metabolismo de la FB_1 en el ser humano. En animales experimentales se absorbe muy poco cuando se administra por vía oral, se elimina rápidamente de la circulación y se recupera sin metabolizar en las heces. La excreción biliar es importante y se excretan pequeñas cantidades en la orina. Se puede degradar a FB_1 parcialmente hidrolizada en el intestino de primates no humanos y en algunos rumiantes. Se retiene una pequeña cantidad en el hígado y el riñón.

1.6 Efectos en los animales y en los sistemas de prueba *in vitro*

La FB_1 es hepatotóxica en todas las especies animales sometidas a prueba, en particular ratones, ratas, équidos, conejos, cerdos y primates no humanos. Con la excepción de los hámsteres sirios, sólo se observa embriotoxicidad o teratogenicidad cuando se produce toxicidad materna o después de ella. Las fumonisinas son nefrotóxicas en cerdos, ratas, ovejas, ratones y conejos. En ratas y conejos se produce toxicidad renal a dosis inferiores a las de la hepatotoxicidad. Se sabe que las fumonisinas producen leucoencefalomalacia equina y síndrome de edema pulmonar porcino, asociados ambos con el consumo de piensos a base de maíz. Es limitada la información sobre las propiedades inmunológicas de la FB_1. Fue hepatocarcinogénica para las ratas machos de una raza y nefrocarcinogénica en otra raza, utilizando la misma dosificación (50 mg/kg de alimentos) y fue hepatocarcinogénica con 50 mg/kg de alimentos en ratones hembra. Parece haber una correlación entre la toxicidad en los órganos y la aparición de cáncer. La FB_1 fue el primer inhibidor específico del metabolismo de los esfingolípidos *de novo* que se descubrió y se está utilizando ampliamente en la actualidad para estudiar su función en la regulación celular. La FB_1 inhibe el crecimiento celular y produce la acumulación de bases esfingoideas libres y la alteración del metabolismo lipídico en animales, plantas y algunas levaduras. No indujo mutaciones génicas en bacterias o síntesis no programada de ADN en hepatocitos primarios de rata, pero sí un aumento de las aberraciones cromosómicas dependiente de la dosis con concentraciones bajas en un estudio sobre los hepatocitos primarios de rata.

1.7 Efectos en el ser humano

No hay registros confirmados de toxicidad aguda de la fumonisina en el ser humano. Los estudios de correlación disponibles procedentes de Transkei (Sudáfrica) parecen indicar una vinculación entre la exposición a la fumonisina en los alimentos y el cáncer de esófago. Esto se ha observado en lugares donde se ha demostrado una exposición relativamente alta a la fumonisina y donde las condiciones ambientales favorecen la acumulación de fumonisina en el maíz, que es el alimento básico. También hay estudios de correlación de China. Sin embargo, no se obtuvo una imagen clara de la relación entre la contaminación bien por fumonisina o bien por *F. verticillioides* y el cáncer de esófago. Debido a la ausencia de datos sobre la exposición a la fumonisina, no se puede llegar a ninguna conclusión a partir de un estudio de casos y testigos de varones en Italia que mostraba una asociación entre el consumo de maíz y el cáncer en la parte superior del aparato gastrointestinal en personas con un elevado consumo de alcohol.

No hay biomarcadores validados para la exposición humana a la FB_1

1.8 Efectos en otros organismos en el laboratorio

La FB_1 inhibe el crecimiento celular y produce acumulación de bases esfingoideas libres y la alteración del metabolismo lipídico en *Saccharomyces cerevisiae.*

La FB_1 es fitotóxica, provoca lesiones en las membranas celulares y reduce la síntesis de clorofila. También altera la biosíntesis de esfingolípidos en las plantas y puede desempeñar una función en la patogenicidad del maíz por las especies de *Fusarium* que producen fumonisina.

2. Evaluación de los riesgos para la salud humana

2.1 Exposición

La exposición humana, demostrada por la presencia de FB_1 en el maíz para consumo humano, es común en todo el mundo. Hay diferencias considerables en el grado de exposición humana entre las diferentes regiones de cultivo de maíz. Esto se pone de manifiesto sobre todo cuando se establece una comparación entre países plenamente desarrollados y en desarrollo. Por ejemplo, aunque puede haber FB_1 en productos de maíz en los Estados Unidos, el Canadá y Europa occidental, el consumo humano de estos productos es pequeño. En algunas partes de Africa, América del Sur y Central y Asia, algunas poblaciones consumen un elevado porcentaje de sus calorías como harina de maíz, cuya contaminación por FB_1 puede ser alta (véase el apéndice 2). El maíz contaminado de forma natural por FB_1 puede estar contaminado simultáneamente por otras toxinas de *F. verticillioides* o *F. proliferatum* o por otras toxinas importantes desde el punto de vista agrícola, en particular el deoxinivalenol, la zearalenona, la aflatoxina y la ocratoxina.

La FB_1 es estable en los métodos de elaboración de alimentos que se utilizan en América del Norte y Europa occidental. El tratamiento del maíz con bases y/o el lavado con agua reduce de manera eficaz las concentraciones de FB_1. Sin embargo, en animales experimentales siguen siendo evidentes su hepatotoxicidad y/o nefrotoxicidad. Se sabe poco acerca de la influencia de las técnicas de elaboración de alimentos utilizadas en el mundo en desarrollo en la FB_1 en los productos de maíz.

2.2 Identificación de peligros

Se ha demostrado la función causal de la exposición a la FB_1 en la leucoencefalomalacia equina. Durante el siglo XIX se produjeron en los Estados Unidos brotes en gran escala de esta enfermedad letal, y también en épocas tan recientes como 1989-1990. Se ha establecido asimismo la función causal de la exposición a la FB_1 en la enfermedad mortal del edema pulmonar porcino. Tal como se observó en hembras preñadas, una exposición baja a la FB_1 es letal para los conejos. Se ha demostrado que la exposición provoca toxicidad renal y hepatotoxicidad en todas las especies animales estudiadas, incluidos los primates no humanos. La exposición a la FB_1 produce hipercolesterolemia en varias especies animales, en particular en primates no humanos. Hay pruebas convincentes de que en las enfermedades animales asociadas con la exposición a la FB_1 se altera el metabolismo lipídico. Es manifiesta la perturbación del metabolismo de los esfingolípidos antes de la toxicidad *in vitro* e *in vivo* o coincidiendo con ella. El uso de fumonisinas como instrumento para estudiar la función de los esfingolípidos ha puesto de manifiesto que éstos se requieren para el crecimiento celular y afectan de varias formas a la señalización de las moléculas, provocando muerte celular apoptótica y necrótica, diferenciación celular y respuestas inmunitarias alteradas. Parece que son factores comunes tras la exposición a la FB_1 la alteración del metabolismo lipídico y los cambios en la actividad y/o la expresión de enzimas fundamentales encargados del funcionamiento normal del ciclo celular. La FB_1 no es tóxica para el desarrollo en ratas, ratones o conejos. A dosis elevadas sin toxicidad materna induce fetotoxicidad en el hámster sirio.

La carcinogenicidad de la FB_1 en roedores varía en función de las especies, las razas y el sexo. El único estudio con ratones $B6C3F_1$ puso de manifiesto que la FB_1 era hepatocarcinogénica para las hembras a 50 mg/kg en los alimentos. En ratas BD IX macho que recibieron alimentos con 50 mg de FB_1/kg durante un período de hasta 26 meses se observó la inducción de carcinomas hepatocelulares primarios y carcinomas colangiales. Se detectaron adenomas y carcinomas de los túbulos renales en ratas F344/N Nctr macho a las que se suministraron 50 mg de FB_1/kg. Parece existir una correlación entre la toxicidad en los órganos y la aparición de cáncer.

El número de estudios de genotoxicidad disponible es limitado. La FB_1 no fue mutagénica en valoraciones bacterianas. En un estudio realizado con células de mamífero *in vitro* no se detectó síntesis de ADN no programado, pero la FB_1 provocó roturas cromosómicas en hepatocitos de rata. En otros estudios se ha puesto de manifiesto que la FB_1 provoca un aumento de la peroxidación de los lípidos *in vivo* e *in vitro*. Es posible que los efectos de la rotura de los cromosomas y la peroxidación de los lípidos tengan una relación causal.

Las concentraciones de FB_1 superiores a 100 mg/kg, notificadas en el maíz de consumo humano en Africa y China, probablemente provoquen leucoencefalomalacia, síndrome de edema pulmonar o cáncer si se administran a caballos, cerdos y ratas o ratones, respectivamente. A pesar de estos casos de exposición humana muy alta, no hay datos confirmados de intoxicación aguda por fumonisina en personas. Los estudios de correlación disponibles del Transkei, Sudáfrica, parecen indicar una relación entre la exposición a la fumonisina a través de los alimentos y el cáncer de esófago. Se han observado índices elevados de cáncer de esófago donde se ha demostrado una exposición relativamente elevada a la fumonisina y donde las condiciones ambientales favorecen la acumulación de fumonisina en el maíz, que es el alimento básico.

En un estudio de casos y testigos realizado con varones en Italia se observó una asociación entre el consumo de maíz y la aparición de cáncer en la parte superior del aparato digestivo, incluido el cáncer de esófago, entre personas habituadas a un consumo de alcohol alto. No había datos sobre la exposición a la fumonisina.

2.3 Evaluación de la respuesta en función de la dosis

La dosis más baja de FB_1 que indujo hepatocarcinomas en animales experimentales fue de 50 mg/kg de alimentos en ratas BD IX macho y en ratones B6C3F₁/Nctr hembra; no se observó inducción de cáncer con 25 ó 15 mg/kg de alimentos, respectivamente. En cada caso, se detectaron indicios de hepatotoxicidad o alteraciones lipídicas con dosis iguales o inferiores en estudios con esas mismas razas de ratas y ratones. La dosis más baja de FB_1 que indujo carcinomas renales en ratas F344/N Nctr macho fue de 50 mg/kg de alimentos; no se observó inducción de cáncer con 15 mg/kg de alimentos. Se produjo apoptosis tubular renal y proliferación celular, así como cambios en los esfingolípidos tisulares y urinarios, con dosis inferiores a las normalmente necesarias para la inducción de cáncer en esos estudios.

No se dispone de datos para evaluar cuantitativamente la relación entre la exposición a la FB_1 y los posibles efectos en el ser humano.

2.4 Caracterización del riesgo

La FB_1 es carcinogénica en ratones y ratas e induce enfermedades letales en cerdos y caballos en concentraciones de exposición a las que están sometidas los seres humanos. El Grupo Especial no estuvo en condiciones de realizar una estimación cuantitativa de los riesgos para la salud humana, pero consideró que se necesitaba con urgencia dicha estimación.

3. Recomendaciones para la protección de la salud humana

a) Se deben establecer límites para la exposición de las personas a través de los alimentos. Se debe prestar particular atención a las poblaciones que consumen un porcentaje elevado de sus calorías como harina de maíz.

b) Se deben adoptar medidas para limitar la exposición a la fumonisina y la contaminación del maíz mediante:

- plantación de cultivos alternativos en zonas donde el maíz no esté bien adaptado;
- obtención de maíz resistente a la podredumbre del grano por *Fusarium*;
- aplicación de mejores prácticas de cultivo;
- separación de los granos mohosos.

c) Se debe aumentar la sensibilización temprana acerca de la posibilidad de contaminación de los alimentos, mejorando la comunicación entre los veterinarios y los funcionarios de salud pública sobre los brotes de micotoxicosis en animales domésticos.

d) Se debe elaborar un método sólido, de bajo costo y sencillo para la detección de la contaminación por fumonisina en el maíz.

www.ingramcontent.com/pod-product-compliance
Lightning Source LLC
Chambersburg PA
CBHW071642210326
41597CB00017B/2081